PSYCHOLOGY THROUGH THE EYES
OF FAITH

D1114022

ALSO BY DAVID G. MYERS

The Human Puzzle: Psychological Research and Christian Belief

The Inflated Self: Human Illusions and the Biblical Call to Hope

Inflation, Poortalk, and the Gospel (*with T. Ludwig, M. Westphal, R. Klay*)

Social Psychology

The Human Connection (*with M. Bolt*)

Psychology

ALSO BY MALCOLM A. JEEVES

Analysis of Structural Learning

Experimental Psychology: An Introduction for Biologists

The Effect of Structural Relations upon Transfer

Thinking in Structures

Free to Be Different

Psychology and Christianity: The View Both Ways

The Scientific Enterprise and Christian Faith

PSYCHOLOGY
THROUGH
THE EYES OF FAITH

David G. Myers
Hope College
Holland, Michigan
and
Malcolm A. Jeeves
University of St. Andrews
Scotland

Christian College Coalition
For Enduring Values

1817

Harper & Row, Publishers, San Francisco

Cambridge, Hagerstown, New York, Philadelphia, Washington
London, Mexico City, São Paulo, Singapore, Sydney

The Christian College Coalition, an association of more than seventy-five colleges, was founded in 1976. The member colleges of the Coalition, affiliated with nearly thirty denominations, find enrichment in the diversity of religious traditions within the Coalition. With main offices in Washington, D.C., the Coalition coordinates programs for regionally accredited liberal arts colleges and universities that offer excellent academic programs and biblical orientation on life and learning. For more information write to:

Christian College Coalition
1776 Massachusetts Avenue, N.W.
Washington, D.C. 20036

Portions of this work originally appeared in *Christianity Today* and the *Church Herald*.

PSYCHOLOGY THROUGH THE EYES OF FAITH. Copyright © 1987 by Christian College Coalition. All rights reserved. Printed in the United States of America. No part of this book may be used or reproduced in any manner whatsoever without written permission except in the case of brief quotations embodied in critical articles and reviews. For information address Harper & Row, Publishers, Inc., 10 East 53rd Street, New York, N.Y. 10022. Published simultaneously in Canada by Fitzhenry & Whiteside, Limited, Toronto.

FIRST EDITION

Library of Congress Cataloging-in-Publication Data

Myers, David G.
  Psychology through the eyes of faith.

  Bibliography: p.
  Includes index.
  1. Christianity—Psychology.   I. Jeeves, Malcolm A.,
1926–   II. Title.
BR110.M94   1987   261.5'15   87-45190
ISBN 0-06-065557-7 (pbk.)

87 88 89 90 91 MPC 10 9 8 7 6 5 4 3 2 1

PSYCHOLOGY TASK FORCE MEMBERS

Dr. David Benner
Wheaton College

Dr. Mark Cosgrove
Taylor University

Dr. Bert Hodges
Gordon College

Dr. Cecil Paul
Eastern Nazarene College

Dr. Mary Vander Goot
Calvin College

SERIES ADVISORY BOARD MEMBERS

Dr. Nicholas Wolterstorff, Editor in Chief
Calvin College
Free University of Amsterdam

Dr. David Benner
Wheaton College

Dr. Richard Bube
Stanford University

Dr. David Allen Hubbard
Fuller Theological Seminary

Dr. Karen Longman
Christian College Coalition

Dr. Ann Paton
Geneva College

Dr. Timothy Smith
Johns Hopkins University

Dr. Dick Wright
Gordon College

CONTENTS

FOREWORD

Suppose that we look with care at contemporary psychology through the eyes of Christian faith. What will we see? What questions will the enterprise and its results lead us to ask? What will we find illuminating? What will we find misguided? At what points will we feel ourselves rightly criticized? The goal of this book is to offer guidance in the answering of these questions.

It is the conviction of the authors of this book, and of those who have sponsored its writing and publication, that although psychology and Christianity are profoundly different, yet they interact. There are some who would deny that. They would insist that Christian conviction has nothing to do with what psychology talks about; science is science and religion is religion and when each sticks to its tasks, ne'er the two shall meet. It is our conviction, by contrast, that Christianity is not just an expression of feelings or values but that it incorporates a body of claims; and that those claims are not just about God but about this world of ours, especially its human beings. Christianity incorporates nothing less than a world view. And about the human beings in that world, psychology speaks.

What inspires us then is a passion for bringing some wholeness, some integrity, into the lives of those of us who are both committed to Christianity and students of psychology. Sometimes the Christian studying psychology will feel tension. When are these felt tensions grounded in real tension and when are they merely apparent? And what is to be done about the real tensions? When should the Christian say to the psychologist, we think you're on the wrong track? When, conversely, should Christians allow the psychologist to convict *them* of being on the wrong track? And where, instead of tension between these two realities, is there illumination of the one by the other? Where does Christianity illuminate what goes on in psychology? Where does psychology illuminate the practices and convictions of Christians?

Of course, given that Christianity and psychology intersect and interact, one can in principle either look at Christianity from within some psychological framework or look at psychology from within Christianity. This book does the latter. This is psychology seen through the eyes of faith.

Let it be added, however, that the two authors, Myers and Jeeves, who function as guides in this enterprise, are by no means outsiders to the field of psychology. Both have deservedly high reputations within the mainstream of contemporary psychology. They are appreciative without being gullible, committed without being naive, critical without being judgmental. They raise profound questions. But also they take bemused note of some of the oddities that turn up along the way.

Not everything gets neatly wrapped up. This book, to say it once again, is for those who want some wholeness in their thought and lives—for those Christians interested in psychology who want some wholeness, and for all others who care to listen in. It offers plenty of suggestions. But the authors don't know the answers to all the questions they raise; and when they don't, they don't pretend they do. So the book is not only guide but invitation—an invitation to engage in the open-ended project of Christian learning.

Naturally there are other Christian psychologists who would place the emphases differently. Myers and Jeeves call our attention to tensions, real tensions, between Christianity and much of what goes on in contemporary psychology. They express their misgivings about some of the values and assumptions implicit in psychology. But it is not on tensions that their emphasis falls. They emphasize the illumination that psychology provides for Christians, and the fact that Christian conviction and psychology results sometimes represent complementary ways of getting at the same features of human beings. Others, without denying the presence of such illumination and such harmony, would have written out of the conviction that the need of the day is for a more radical Christian critique of contemporary psychology. So it should not be supposed that this book represents the consensus of emphasis among Christian psychologists. There is no such consensus. It represents, rather, one position within the debate; and Myers and Jeeves explain some of the other positions. Thus to read the book is to enter a dialogue among Christian psychologists. In that way too, it is meant not as closure but as invitation.

NICHOLAS WOLTERSTORFF
EDITOR IN CHIEF

Professor of Philosophy
Calvin College
Free University of Amsterdam

PREFACE

This book is about the relationship between what psychologists are discovering and what Christians believe. Unlike most other books on psychology and religion, this one is not primarily concerned with what psychologists have said about religion, with the speculations of personality theorists, or with a Christian approach to counseling. Rather, it identifies major insights regarding human nature that college and university students will encounter in a basic psychology course and ponders how the resulting human image connects with Christian belief.

Readers will, we hope, include all sorts of intellectually curious people, including students eager for a Christian perspective on some of the provocative new developments in psychology. With students' needs in mind, we have written short essays on many of the big issues in psychology and religion and arranged them to correspond with the sequence of topics often studied in a first course on psychology.

We gratefully acknowledge the advice and encouragement of dozens of psychology professors from the member colleges of the Christian College Coalition. We especially benefited from stimulating discussions with the Coalition's psychology advisory committee—David Benner, Mark Cosgrove, Bert Hodges, Cecil Paul, and Mary Vander Goot. Philosopher Nicholas Wolterstorff, editor-in-chief for this pioneering series of books, and Christian College Coalition executives John Dellenback and Karen Longman, provided much warm-hearted encouragement from beginning to end.

Our thanks also go to the University of St. Andrews for hosting

David Myers during the drafting phase, to Hope College for its sabbatical support of the project, to Julia Zuwerink for her editorial assistance, and to Phyllis Vandervelde for her professional excellence in producing manuscript drafts.

Chapter 1

LESSONS FROM THE PAST: SCIENCE AND CHRISTIAN FAITH

I am free, I am bound to nobody's word, except to those inspired by God; if I oppose these in the least degree, I beseech God to forgive me my audacity of judgment, as I have been moved not so much by longing for some opinion of my own as by love for the freedom of science.

NATHANIEL CARPENTER,
PHILOSOPHIA LIBERA, 1622

Over the last century—psychology's first century—definitions of the field have varied. For its first forty years psychology was, as William James declared in his pioneering 1890 text, *The Principles of Psychology*, "the science of mental life." During the next forty years, from the 1920s into the 1960s, it was the science of behavior. Today's textbooks commonly synthesize this history by defining the field as the science of behavior and mental processes.

Note what all these definitions have in common: that psychology is a *science*. So, in considering the question "What is the relationship between Christian faith and psychology?" let's begin with the history of relations between faith and science.

Asked, "What is the relationship between faith and science?" many people—Christians and non-Christians alike—answer, "Conflict." They think of Galileo, condemned for questioning the church's conviction that the sun revolves around a stationary earth. They think of the reaction against Darwin's ideas at the Scopes trial and among today's antievolutionists. They think of the encroachment of natural explanations of disease, of earthquakes and storms, and of human behavior—realms once reserved for supernatural explanation. If religious and scientific explanations occupy opposite ends of a teeter-totter, than as one goes up, the other must come down.

Contrary to this popular view that religion and science are antagonistic, many intellectual historians argue that the development of modern science was supported by Hebraic-Christian ideas that during the seventeenth century came to replace ancient assumptions about God and nature. If, as had often been supposed, nature is sacred, then we ought not tamper or experiment with it. If, however, nature is not an aspect of God, but rather is God's intelligible creation—a work to be enjoyed and managed—then by all means let us explore this handiwork. If we wish to discover its order, let us observe and experiment, believing that whatever God found worth creating we can find worth studying. Moreover, let us do so freely, knowing that our ultimate allegiance is not to any human authority or doctrine, but to God alone.

It was this biblical view of God and nature that encouraged several of the founders of modern science (among them Blaise Pascal, Francis Bacon, Isaac Newton, and even Galileo) and many of the founders of American colleges, ninety percent of which were church-founded at the time of the Civil War. Whether searching for truth in the book of God's word or the book of God's creative works, these scientific pioneers viewed themselves in God's service. Believing that humans, too, were finite creatures of God, not extensions of God, they did not depend solely on intuition and reason but also observation. They assumed that we cannot find the whole truth merely by searching our minds—for there is not enough there—nor merely by guessing or making up stories.

Instead, their ideal was humbly to submit their ideas to the test, knowing that if nature did not conform to these then so much the worse for their ideas. Having dominion over nature meant not to force nature into their doctrinal categories, but rather first to understand it, then to adapt their conceptions to what their observations and experiments revealed. If their data told them that the earth was not stationary, then they must abandon the notion that heavenly bodies circled the earth. Reason, they believed, must be aided by observation and experiment in matters of science, and by spiritual revelation in matters of faith.

This Hebraic-Christian foundation for scientific pursuits applies also to the scientific study of human nature, because humans, too, are part of the created order. In the Hebrew Scriptures, humans are created by God "of dust from the ground." Thus after gazing at the heavens the psalmist could wonder, "What is man that thou art mindful of him?" Yet this human creature was a special creation, a majestic summit of God's creative activity of whom the psalmist could in the next breadth rhapsodize, "Thou hast made him little less than God, and dost crown him with glory and honor. Thou hast given him dominion over the works of thy hands."

So what is the relationship between science and faith? There are points of tension, to be sure, some of which will be identified in chapters to come. But it is also true that the Christian idea of a sovereign Creator has nourished the scientific spirit. Moreover, the scientific enterprise may properly be seen as one way of fulfilling the call to stewardship, of obeying the command to manage and care for the created order.

And consider: we humans are firmly placed within the natural order. As God's creatures, we are dependent upon God's sustaining power, moment by moment. This dependence upon and allegiance to God alone frees us from bondage to anybody's word, except to those found in God's books. We are freed even to investigate that most marvelous wonder of nature—human nature. To paraphrase the historian of science R. Hooykaas, what the Bible urges upon us is a complete transformation in our relations to God and our fellow creatures, and to the world that God has made. This transformation means a liberation from old superstitious bonds and from any kind of idolatry, including the idols of common opinion and official doctrine. We who have been touched by the Spirit may respect human authorities in church, state, or science, but we will not be so deeply impressed by them that we give up our independence. Our liberation implies also a new obedience by which we must be willing to submit all our prejudices and all our prior criteria of reasonableness to the test of divine revelation, including the reality of the universe around us.

For Further Reading

Hooykaas, R. *Religion and the Rise of Modern Sience*. Grand Rapids, Mich.: Eerdmans, 1972.
    Describes and documents the religious roots of contemporary science.

Jeeves, M. A. *The Scientific Enterprise and Christian Faith*. London: Tyndale Press, 1968; Downers Grove, Ill.: InterVarsity, 1971.
    Christians in various scientific fields explain how their faith relates to their science.

Lindberg, D. C., and R. L. Numbers, eds. *God and Nature: Historical Essays on the Encounter Between Christianity and Science*. Berkeley and Los Angeles: University of California Press, 1986. Eighteen scholarly essays explore the progressive and regressive influences of Christianity upon the development of modern science.

Chapter 2

LEVELS OF EXPLANATION

Reality is a multi-layered unity. I can perceive another person as an aggregation of atoms, an open biochemical system in interaction with the environment, a specimen of *homo sapiens,* an object of beauty, someone whose needs deserve my respect and compassion, a brother for whom Christ died. All are true and all mysteriously coinhere in that one person.

JOHN POLKINGHORNE,
*ONE WORLD: THE INTERACTION OF
SCIENCE AND THEOLOGY,* 1986

Scan any textbook of psychology and immediately you will be struck by an incredible variety of approaches. At the back of the book one typically reads of social-psychological investigations of how people are influenced by their groups. Near the middle of the book, one finds the work of those who study learning, thinking, and memory. But these topics can also be analyzed in terms of their biological components. Thus near the beginning of the book one is introduced to neuropsychological principles of brain organization and nerve transmission, and to the chemical messenger system by which nerve cells communicate.

You might say that each of us is a complex system that is part of a larger social system, but also that each of us is composed of smaller systems, such as our nervous system and body organs, which are composed of still smaller and smaller systems—cells, biochemicals, atoms, and so forth. Any given phenomenon, such as thinking, can be viewed from the perspective of almost any one of these systems—from the social influences on thinking to biochemical influences. The variety of possible perspectives, or *levels of analysis* as they are also called, requires that we choose which level we wish to operate from. Each level entails its own questions and its own methods. Each provides a valuable way of looking at behavior, yet each by itself is incomplete. Thus each level complements the others; with all the perspectives we have a more complete view of our subject than provided by any one perspective.

Take memory. Neuropsychologists study the chemical codes and neural networks in which information is stored and the importance of particular parts of the brain for particular kinds of memory. Cognitive psychologists study memory in nonphysical terms, as a partly automatic and partly effortful process of encoding, storing, and retrieving information. Social psychologists study the effects of our moods and social experiences upon our recall of past events.

Psychologists working at each of these levels accept that even if their explanations were to become complete in their own terms this would *not* invalidate or preempt the other levels of explanation. The neuropsychological perspective, for example, is extremely valuable for certain purposes, but is not so valuable for understanding, say, social relations.

It's like viewing a masterpiece painting. If you stand right up against it you will understand better how the paint was applied, but you will miss completely the subject and impact of the painting as a whole. To say the painting is "nothing but" or "reducible to" blobs of paint may at one level be true, but it misses the beauty and meaning that can be seen if one steps back and views the painting as a whole. To consider a phone caller's voice as reducible to electrical impulses on the phone line is extremely useful for some scientific purposes, but if you view it as nothing more you will miss its message. For the electrical engineer's purposes the message is irrelevant, much as God's activity is, in one sense, superfluous to a scientific account of the mechanisms by which God's creation operates. Yet for the sorts of questions that Leo Tolstoy agonized over—"Why should I live? Why should I do anything? Is there in life any purpose which the inevitable death which awaits me does not undo and destroy?"—we find the "God hypothesis," the perspective of faith, exceedingly helpful.

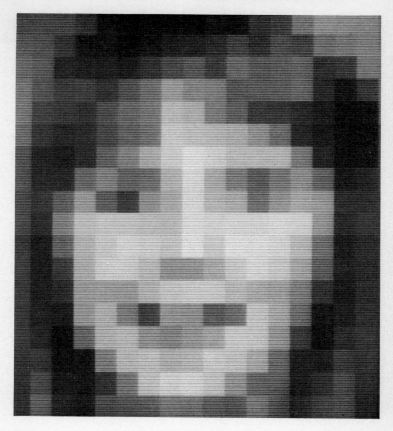

Figure 1: Levels of analysis. What do you see? On close inspection, this image appears to be "nothing but" its computer-produced blocks. Viewed from a different level of analysis—from ten feet or more away—we gain a more holistic perspective and see what it truly is: a photo of ten-year-old Laura Myers. Note that each perspective is valid. Look only from a distance and you will never see what the photo is made of. Look only close up and you will miss the whole picture. (Portrait courtesy of Cecil W. Thomas, Ph.D., Biomedical Engineering Department, and Grover C. Gilmore, Ph.D., and Fred L. Royer, Ph.D., Psychology Department, Case Western Reserve University, Cleveland, Ohio.)

Integrative Explanation

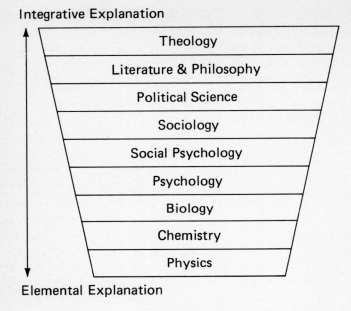

Elemental Explanation

Figure 2: Partial hierarchy of disciplines. The disciplines range from basic sciences that study nature's building blocks up to more integrative disciplines that study whole complex systems. Successful explanation of human functioning at one level need not invalidate explanation at other levels.

What is true of psychology is also true of the other academic disciplines, each of which provides a perspective from which we can study nature and our place in it. These range from the scientific fields that study the most elementary building blocks of nature up to philosophy and theology, which address some of life's global questions.

Which of the various perspectives is pertinent all depends on what you want to talk about. Take romantic love, for example. A physiologist might describe love as a state of arousal. A social psychologist would examine how various characteristics and conditions—good looks, similarity of the partners, sheer repeated exposure to one another—enhance the emotion of love. A poet would express the sublime experience that love can sometimes be. A theologian might describe love as the God-given goal of human relationships. Since an event, like love, can often be described simultaneously at various levels, we need not assume that one level is causing the other by supposing, for example, that a brain state is causing the emotion of love or that the emotion is causing the brain state. The emotional and physiological perspectives are simply two ways of looking at the same event.

Nature is, to be sure, all of a piece. For convenience, we necessarily view it as multilayered, but it is actually a seamless unity. Thus the different ways of looking at a phenomenon like romantic love or belief or consciousness can sometimes be correlated, enabling us to build bridges between different perspectives. Attempts at building bridges between religion and the human sciences have sometimes proceeded smoothly. A religious explanation of the incest taboo (in terms of divine will or a moral absolute) is nicely complemented by biological explanation (in terms of the genetic penalty that offspring pay for inbreeding) and sociological explanation (in terms of preserving the marital and family units). At other times the bridge-building efforts extending from both sides seem not to connect in the middle, as when a conviction that God performs miracles in answer to prayer is met with scientific skepticism and psychological explanations of how people form illusory beliefs. To agree, as we have, that religious and scientific levels of explanation can be complementary does not mean there is never conflict or that any unsupported idea is to be welcomed as truth. It just means that different types of explanation may actually fit coherently together. In God's world, all truth is one.

So we arrive at a simple but basic point that resolves a good deal of fruitless debate over whether the religious or the psychological account of human nature is preferable: different levels of explanation can be complimentary. The methods of psychology are appropriate, and appropriate *only*, for their own purposes. Psychological explanation has provided satisfying answers to many important questions regarding why people think, feel, and act as they do. But it does not even pretend to answer life's ultimate questions. Let us therefore celebrate and use psychology for what it offers us, never forgetting what it cannot accomplish.

For Further Reading

MacKay, D. M. *Human Science and Human Dignity*. Downers Grove, Ill.: InterVarsity Press, 1979.
A brilliant British brain scientist explains why those holding to a biblical view of human nature can welcome the science of human behavior.

Chapter 3

SHOULD THERE BE A CHRISTIAN PSYCHOLOGY?

It is doubtless impossible to approach any human problem with a mind free from bias.

SIMONE DE BEAUVOIR,
THE SECOND SEX, 1953

Simple questions sometimes have not-so-simple answers. This chapter's title is one such question. To see what it might mean to create a Christian psychology or even a Christian perspective on psychology, consider what psychology is. Opening an introductory psychology textbook, we discover that psychology is not a single unified discipline but a federation of subdisciplines. Some psychologists study the responses of single cells, others the behavior of thousands of people in crowds. At some points psychology therefore borders on physiology, at others on sociology.

For many of us, this interdisciplinary breadth is what makes the field so exciting. But it also means that the relevance of faith for psychology may vary, depending on whether one is studying brain mechanisms or, say, personality and counseling methods. Christians who are researching the intricate mechanisms of vision, memory, or language are likely to see their science and their faith as complementary: science explores the natural processes underlying such phenomena while faith helps one grasp the significance of the whole human system. This view—which we endorsed in Chapter 2—views science as offering an important but *limited* perspective on human nature. Other Christians, especially those who study personality or practice therapy with troubled people, are more struck by the ways in which psychologists are influenced by their own values and assumptions. They note that one's values and assumptions—one's whole world view—shape one's topics of study, methods of inquiry, interpretations of data, and applications. Thus they urge Christian psychologists to be alert to psychology's hidden values and assumptions and to Christianize these. To appreciate the rationale behind this "Christianizer" view of psychology and faith, let's consider some examples of how psychologists are indeed influenced by their values and assumptions.

Hidden Values and Assumptions

It is a point worth emphasizing, one that students of any scientific discipline, including psychology, ought never forget: science involves more than impersonal, objective, pure facts. Observations are organized on the basis of our experience and interests. We decide what to pay attention to and what to ignore. This subjective element of scientific exploration is even larger in the human than the physical sciences. Thus *psychologists' personal values penetrate their work in several subtle and not-so-subtle ways.*

Values first of all influence the people that have been attracted to psychology. Surveys of American psychologists during the 1960s and 1970s revealed them to be among the most irreligous academics. A third of them denied the existence of God (nearly ten times the proportion of other Americans) and only a third described themselves as even moderately religious. We wonder: do psychologists, like so many laypeople, tend to see the psychological account of human nature as competing with and elbowing out the religious account?

Values influence psychologists' choice of research topics and their ethical standards in conducting research. Our interests in topics such as aggression, sex-role socialization, and smoking prevention are motivated by our personal concerns.

Values also have more subtle effects. There is a growing awareness among both scientists and philosophers that science is not so purely objective as often presumed. Scientists do not merely read what is out there in the book of nature. Rather, they decide what methods to explore it with, what to observe, and how to interpret their findings. As we will demonstrate in a later chapter, our preconceptions act as a flashlight, riveting our attention on selected aspects of nature. Politically conservative psychologists have therefore tended to explain intelligence in terms of heredity and to do the type of research that supports that view; politically liberal psychologists have tended to favor environmental explanations of intelligence variations and to do the type of research that points to environmental influences.

Values further influence our conceptions of mental and sexual health, of self-actualization and fulfillment, of how best to rear children. Is it better to express and act on one's feelings or to exhibit self-control? To seek joy in the here and now or to endure stress now for the sake of future achievement? Little wonder that in one survey, 425 mental health professionals were almost equally divided on whether it was desirable for people to "become self-sacrificing and unselfish."

What do you think: Should children be trained from birth "in regularity of feeding, sleeping, elimination" so that they might learn that they are "part of the world bigger than their own desires," as the 1938 U.S. Government pamphlet *Infant Care* advised? Or when fussy should they (as the 1942 edition of the same pamphlet advised) immediately be offered "the milk, but also the warmth, the sense of being held firmly" to help them become trusting of the world rather than withdrawn and fearful? The answer depends partly on what we understand to be the effects of parenting style; but it also depends on whether one places a higher value on self-control and respect for authority as did the 1938 authors, or on security and independence as did the 1942 authors.

Values even influence our psychological terminology: whether we label those who say only nice things about themselves on personality inventories as having "high self-esteem" or as "defensive"; whether we describe those who favor their own racial and national groups as "ethnocentric" or as exhibiting strong "group pride";

whether we view a persuasive message as "propaganda" or "education."

Psychologists also are subtly affected by their philosophical and cultural assumptions. Consider, for example, Lawrence Kohlberg's influential theory of moral development, a theory that underlies some modern curricula for moral and values education in public schools. Kohlberg contended that children develop morally as their *thinking* proceeds through a sequence of stages, from a "preconventional" morality of pure self-interest, to a "conventional" morality concerned with gaining others' approval or doing one's duty, to (in some "mature" people) a "postconventional" morality of self-chosen principles. Critics question Kohlberg's assumption that morality is more a matter of thinking than acting, and they even more strongly question the humanistic individualism of his assumption that the "highest" or most mature moral stage is exhibited by those who make moral judgments in accord with their self-chosen convictions. Critic Richard Shweder contends that Kohlberg's moral ideal is the view "of an articulate liberal secular humanist" masquerading as psychological truth. Critic Carol Gilligan argues that Kohlberg's ideas are those of the typical Western male; for women, she believes, moral maturity is not so much a matter of abstract ethical principles as of responsible, committed relationships.

So hidden values and assumptions penetrate psychology. They influence psychologists to construct, confirm, and label concepts that support their presuppositions.

Responses to Psychology's Hidden Values and Assumptions

Psychology's critics have sensitized psychologists to the subtle influences of assumed preferences and beliefs. Marxist critics have sensitized us to capitalist assumptions in psychology, feminist critics to implicit masculine values, Christian critics to secular presuppositions. Should we therefore, as some have argued, replace science that aims to be value-free with a science that expresses one's values and assumptions? (Some have called for the establishment of a Marxist psychology, others for a feminist psychology, others for a Christian psychology.)

Some Christian psychologists answer no. These psychologists tend to be Christians who participate in mainstream psychological science; often they do so with a sense of Christian vocation, recognizing both their own limits and the limits of their discipline. One such person, British neuropsychologist Donald MacKay, worried about those who are eager to inject an ideology, even a Christian one, into psychology. He argued that the Christian psychologist's obligation

is to "tell it like it is," knowing that the Author is at our elbow, a silent judge of the accuracy with which we claim to describe the world He has created. In this sense our goal is objective, value-free knowledge. If our limitations, both intellectual and moral, predictably limit our achievement of this ideal, this is something not to be gloried in but to be acknowledged in a spirit of repentance. Any idea that it could justify a dismissal of the ideal of value-free knowledge as a "myth" would be as irrational—and as irreligious—as to dismiss the idea of *righteousness* as a "myth" on the grounds that we can never perfectly attain that.

For MacKay, ourselves, and others, a Christian psychology is one that is faithful to reality. If God has written the book of nature, it becomes our calling to read it as clearly as we can, remembering that we are humble stewards of the creation, answerable to the giver of all data for the accuracy of our observations. Indeed, it is precisely because all our ideas are vulnerable to error and bias—including our biblical and theological interpretations as well as our scientific concepts—that we must be wary of absolutizing any of our theological or scientific ideas. As the Reformation motto *ever-reforming* suggests, our religious and scientific ideas are mere approximations of truth that always are subject to test, challenge, and revision. Believing that both the natural and biblical data reveal God's truth, we can allow scientific and theological perspectives to challenge and inform each other. But we do so remembering that science and theology operate at different levels of explanation and mindful of the tentative nature of any scientific or theological theory.

There is an additional reason why the Bible does not give us a completed psychology and why we therefore need psychological science. The Scriptures must embody truth not just for us in our twentieth century age but for all people, past, present, and future. For the very same reason, noted C. S. Lewis,

Christianity has not, and does not profess to have, a detailed political programme for applying "Do as you would be done by" to a particular society at a particular moment. It could not have. It is meant for all men at all times and the particular programme which suited one place or time would not suit another. And, anyhow, that is not how Christianity works. When it tells you to feed the hungry it does not give you lessons in cookery. When it tells you to read the Scriptures it does not give you lessons in Hebrew and Greek, or even in English grammar. It was never intended to replace or supersede the ordinary human arts and sciences: it is rather a director which will set them all to the right jobs, and a source of energy which will give them all new life, if only they will put themselves at its disposal.

To repeat, we agree that our values and assumptions cloud the spectacles through which we view reality, but also that our calling is to clean the spectacles through careful scientific and biblical scholarship. Other Christian psychologists, whom philosopher C. Stephan Evans has called the Christianizers of psychology, remind us that psychologists *never* approach their subject completely free of prior beliefs and prejudices. Thus if Christian psychologists are to be fully serious both as scholars and Christians, they must not wall their scientific and religious levels of understanding off from each other. Instead, they should allow the content of their faith to inform their psychology (and vice versa), much as they also allow their faith to inform their social awareness, politics, and personal relationships. For example, rather than uncritically accepting Kohlberg's theory of moral development, Christian developmental psychologists might instead want to construct a theory that is rooted in an explicitly Christian understanding of morality. If this new theory makes testable predictions, it can then be subject to testing along with competing theories.

We generally favor the first of these two Christian responses to psychology, the view that psychological science offers a limited but useful perspective on human nature that complements the perspective of faith. Some chapters that follow therefore describe striking parallels between what researchers are concluding and what Christians believe. Other chapters note how psychological findings can be applied to the concerns of the church—to preaching, prayer, and the quest for faith and happiness. But as this chapter has emphasized, we agree with the Christianizers of psychology that it can matter enormously whether one views human nature through the eyes of faith. Thus still other chapters will look at psychology critically, by calling attention to hidden values and assumptions in psychologists' writings on giftedness, sexuality, and therapy. Through this mixture of Christian criticism, Christian application, and Christian parallels to psychology, we will sample the various ways in which Christians are integrating psychology and faith.

For Further Reading

Evans, C. S. *Preserving the Person.* Grand Rapids, Mich.: Zondervan, 1982.
> Describes and illustrates various approaches that Christians have taken to the human sciences, including the "limiter" and "Christianizer" views briefly described here.

Jones, S. L., ed. *Psychology and the Christian Faith: An Introductory Reader.* Grand Rapids, Mich.: Baker, 1986.
> Jones's helpful summary of perspectives on psychology and faith sets the stage for illustrative chapters by experts in psychology's subspecialties. An especially fine book for those interested in Christian perspectives on personality and psychotherapy.

Van Leeuwen, M. S. *The Person in Psychology: A Contemporary Christian Appraisal.* Grand Rapids, Mich.: Eerdmans, 1985.
> A wide-ranging Christian critique of the discipline of psychology; Van Leeuwen questions prevailing assumptions and proposes a reshaping of traditional scientific methods.

PART 2 / Biological Bases of Behavior

Chapter 4

THE BRAIN-MIND CONNECTION

Fundamental changes in our view of the human brain cannot but have profound effects on our view of ourselves and the world.

DAVID H. HUBEL,
"THE BRAIN," 1979

On June 17, 1783, the famous English author Dr. Samuel Johnson awoke around 3 A.M. and to his surprise and horror found he could not speak. To test his mind, he attempted to compose a prayer in Latin verse and succeeded. Thus reassured, he next tried to loosen his powers of speech by drinking wine, but this only put him back to sleep. When he awoke the next morning he found that he still could not speak, yet he could write and could understand what others said.

What sort of disorder would disrupt speech yet allow one to think, read, write, and listen? Johnson summoned his physicians, who diagnosed a disturbance of the vocal apparatus and prescribed a treatment of blisters on each side of the throat. Sure enough, within a few days his speech began to return, leaving only a slight impediment at the time of his death late the following year.

The ignorance of Johnson's doctors regarding the localization of different aspects of language in the brain was mild compared to that of their predecessors. For many centuries people debated whether the mind was located in the heart, as Aristotle argued in the fourth century B.C., or in the brain, as Hippocrates had guessed. The second-century anatomist Galen, whose views prevailed until the sixteenth century, favored Hippocrates's view, although he mis-located the mind in the brain's fluid-filled ventricles. The early nineteenth-century German physician Franz Gall recognized that various brain regions have specific functions, but he guessed wrongly what they were. By 1865 a French physician, Paul Broca, reported that damage to a specific area on the left side of the brain would produce the speech difficulty that Samuel Johnson suffered (apparently as a result of a mild stroke). And by the 1970s the American neuroscientist Norman Geschwind had assembled more clues into the sequence of brain activities that enables a person to use language. Depending on where in the brain the sequence is broken, a different language disorder (of speaking, reading, or un-derstanding) occurs.

What is true of our understanding of the relation between brain activity and language is true of the brain-mind relation in general: *every new advance in the flourishing field of neuropsychology tight-ens the apparent links between brain and mind*. Even so specific a mental function as the ability to recognize a face is being localized to specific brain regions (principally the lower right side of the brain). In work with monkeys, neuropsychologists have detected specific cells that buzz with activity in response to a specific face or to a specific type of perceived body movement. In humans, detectable brain activity is now known to coincide with and even precede by a fraction of a second the instant at which a person consciously decides to perform an action, such as lifting a finger.

As research accumulates, the link also tightens between brain and personality. We now know that particular types of brain damage have predictable effects on thoughts and emotions, and that ma-nipulating a person's brain can manipulate the person's mind, moods, and motives. And we are learning how abnormalities in the brain's chemical messengers—its neurotransmitters—are in-volved in psychological disorders such as depression and schizo-phrenia. With such findings comes hope that alterations in brain chemistry (through drugs, transplants of brain tissue, or dietary changes) may alleviate emotional suffering.

The Nobel Prize–winning neuroscientist David Hubel is right to suppose that such "fundamental changes in our view of the human brain cannot but have profound effects on our view of ourselves and the world." Scientific advances shape our assumptions about reality. Few assumptions are more fundamental than those involved in the perennial mind-body problem: How does the mind relate to the body?

For centuries the mind-body relationship has puzzled philosophers and scientists. On the one hand, brain activity is, as we have illustrated, tightly linked with mental activity. On the other hand, the mind directs bodily activity: embarrassed, we blush.

One therefore wonders: What is the mind? Is it something immaterial? Does it exist apart from the material brain? (If so, how does it affect the brain?) Or is the mind a manifestation of brain activity?

The first view, *dualism*, presumes that the mind and body are two distinct entities—the mind nonphysical, the body physical—but entities that somehow manage to interact with each other. The ancient Greeks saw mind and body as rider and horse. The Roman philosopher Seneca, referred in his *Morals* to our bodies as our luggage—something we carry around with us. Descartes in the seventeenth century assumed, "I am . . . lodged in my body as a pilot in a vessel." More recently, the neurosurgeon Wilder Penfield argued, "The mind seems to act independently of the brain in the same sense that a programmer acts independently of his computer."

Most brain scientists find this dualistic view hard to accept, partly because it assumes something immaterial that science, dealing only with natural phenomena, can't have any knowledge of. They instead generally favor some form of *monism*, which assumes that mind and body are one. Thus psychologist Donald Hebb could say that however implausible it may be to say that consciousness "consists of brain activity . . . it nevertheless begins to look very much as though the proposition is true."

In the monistic view, which we tend to favor, talking about the brain and talking about the mind are simply two levels of describing the same events. The relationship of brain and mind is therefore something like the relation of the electrical events in a telephone line to what the speaker is saying. The speaker's words are not a ghostly extra that travels separate from the electrical events, but rather a different way of understanding the same events. The electrical transmission and the corresponding words, the brain activity and the corresponding thoughts, are simply two levels at which the same events can be understood. Mortimer Mishkin, neuropsychology chief at the National Institute of Mental Health and president of the Society for Neuroscience, sums up the contemporary view: "The living body and its brain, on the one hand, and behavior and the mind, on the other, are indissoluble."

But how does consciousness arise from brain activity? Somewhere near the top of our list of the great wonders of the world is the emergence of mind from the unimaginably complex interaction of the brain's subsystems. So far as we can tell, mind is not an extra entity that occupies the brain. As Nobel laureate psychologist Roger Sperry emphasized "Everything in science to date seems to indicate that conscious awareness is a property of the living brain and inseparable from it." Yet there it is—our memories, our wishes, our creative ideas, our moment-to-moment awareness—somehow arising from the coordinated activity of billions of nerve cells, each of which communicates with hundreds or thousands of other nerve cells.

An analogy may help us see that the properties of a whole system, such as the brain-mind system, may be united with, yet not be reducible to, its physical parts. Another of the world's wonders is the behavior of the social insects—the ants, the bees, the termites. An ant colony, for example, is a sort of intelligent organism. It "knows" how to grow, how to move, how to build. This intelligence is not reducible to the individual ants; a solitary ant, with only a few neurons strung together, is a witless, thoughtless creature. Yet from the interactions of a dense mass of thousands of ants a collective intelligence somehow emerges. There is nothing extra plugged into the ants to create this intelligence, yet to look no further than the individual ants would be to miss the miracle of the living colony.

Likewise, to stop with the story of the brain cells would be to miss the miracle of human experience. The human part of you and me is not a ghost in a body but rather the whole unified system of brain and mind. Our human experiences of pain and pleasure, of self-awareness and abstract thought, emerge from brain activity yet can be understood at their own level. We may indeed have been created from dust, over eons of time, but the end result is a priceless creature, one rich with potentials beyond our imagining.

For Further Reading

Cosgrove, M. *The Amazing Body Human: God's Design for Personhood.* Grand Rapids, Mich.: Baker, 1987.
> Explores the fascinating differences between human and animal bodies and how those differences, including brain differences, relate to personhood. For more comparisons of humans with other animals see R. L. Kotesky, *Psychology from a Christian Perpective.* Nashville: Abingdon Press, 1980.

Jones, D. G. *Our Fragile Brains: A Christian Perspective on Brain Research.* Downers Grove, Ill.: InterVarsity, 1980.
> Summarizes what we know about the brain and the implications of brain research for a Christian view of persons.

Siegel, R. K. "The Psychology of Life After Death," *American Psychologist* 35 (911–31).
> Do out-of-body "near death" experiences provide evidence for a dualistic view of human nature? Ronald Siegel, a researcher of drug-induced hallucinations, analyzes reports of near-death experiences and explains how such experiences are manufactured by the brain under stress.

Chapter 5

BIBLICAL IMAGES OF HUMAN NATURE

Does not death mean that the body comes to exist by itself, separated from the soul, and that the soul exists by herself, separated from the body? What is death but that?

SOCRATES, IN
PLATO'S *PHAEDO,* FOURTH
CENTURY B.C.

For the Hebrew, man is a unity, and that unity is the body as a complex of parts, drawing their life and activity from a breath-soul, which has no existence apart from the body.

H. WHEELER ROBINSON,
"HEBREW PSYCHOLOGY," 1925

What is our essential nature? Are we a dualism of body and soul, as Socrates and Plato believed, or a psychophysical unity, as H. Wheeler Robinson suggests is the Old Testament view? Take a little survey of Christian laypeople and you will surely find most people agreeing with Plato: we are made up of *two* realities, body and soul. Or so most of us have been taught since childhood. One book of Christian doctrine for children explains:

Maybe you have been to a funeral. You've seen the dead body. That's buried in the ground. But the inside part of you, the part that thinks and feels, that's the part that lives forever. This is the part of us that would go to hell.

As Socrates' discourse on death hints, the idea of an immortal soul arises not from the Bible but from Greek thought. In the end, Plato records that Socrates lived his own teaching by drinking the poison hemlock in the serene conviction that his immortal soul would now find release from its bodily prison. For Socrates and

Plato, bodily death was a welcome liberation. Indeed, it was actually *not dying*.

In the centuries after Christ, theologians combined this Greek doctrine of the immortal soul with biblical images of human nature. When Origen, a third-century platonic philosopher, became the father of theology, he built into Christian doctrine Plato's idea of the soul. In the early fifth century, Augustine thought Plato to be the most bright in all of philosophy. And in the sixteenth century, John Calvin, who was heavily influenced by both Plato and Augustine, declared that Plato alone "rightly affirmed" the immortal soul that "lies hidden in man separate from body."

It has been one of the tasks of twentieth-century biblical scholarship to disentangle the biblical images of human nature from those of Greek philosophy. Discerning the biblical picture of the person is no simple matter, for the Bible is actually not one book but a library of sixty-six books, written over some fifteen hundred years, in three languages, and under varying historical circumstances. Not surprisingly, then, the meanings of the same words within the Hebrew Old Testament and within the Greek New Testament may shift. Thus it becomes important to know what the biblical writers were saying when they described their experiences. One mistake is to interpret their everyday language descriptions as modern, scientific statements. When the writer of Ecclesiasties (1:5) noted that "The sun rises and the sun goes down," the church of Galileo's day understood this to be a scriptural proclamation of a stationary earth encircled by the sun. Because the writer of Genesis (1:16) described the moon as a "light to rule the night" they also could not accept Galileo's conclusion that the moon shone by reflected light. When he invited them to look through his telescope and see the shadows of the craters for themselves, they declined and dismissed his observations as delusions of the devil.

It is similarly possible to misinterpret the Bible's human images. It is important to remember that the Bible is written for all people of all times. As such, it does not intend to offer a precise psychology, and certainly not one in the language of the late twentieth century. Moreover, it is a book for living, not a book of science. It is not biological but biographical. It is about the acts of God in the lives of people throughout history.

Bearing these cautions in mind, what general understanding of human nature emerges from the library of Scriptures? Let's consider some conclusions reached by scholars who have devoted their lives to exploring the whole of Scripture.

Old Testament Images

The Hebrew Scriptures hold in delicate balance the image of humans as the majestic summit of God's creative activity, uniquely made in God's own image, and yet also as very much a part of the creation, formed of the dust of the ground. What scientists have come to understand as the physical underpinnings of mind and emotion was assumed in Hebrew thought. The details of Hebrew psychology differ from the details of contemporary science, but on one fundamental point the two traditions agree: mind and emotion are inextricably linked with body. The people of the Old Testament think with their hearts, feel with their bowels, and their flesh longs for God. Of the body organs the heart (*leb*, in Hebrew) is the most important. In its 851 Old Testament uses, it variously denotes the whole personality, the emotions, or the intellect and will. It is, in these different ways, the center of life, which is called to hear and respond to the word of God.

What then is the biblical "soul"? In Old Testament Hebrew it is *nephesh*, which in its 755 varied uses first of all means physically alive, a living creature, a tangible material person. Out of the dust God formed a person who "became a living *nephesh*" (Gen. 2:7). Contrary to what most people suppose, this *nephesh* is most assuredly not Plato's immortal soul. Our *nephesh*, our whole living being, is said to terminate at death. Moreover, animals as well as humans have *nephesh*. Thus the Old Testament scholar Walther Eichrodt laments, "The unhappy rendering of the term by 'soul' opened the door from the start to the Greek beliefs concerning the soul." The Hebrew sense of *nephesh* is more like the soul we have in mind when we say "there wasn't a soul (person) in the room," or "I love you from the depths of my soul (being)." In the Hebrew view, we do not have *nephesh* (a soul), we *are nephesh* (living beings). We are living beings who were created to share the divine image and to live as obedient stewards of the earth.

New Testament Images

Like the Old Testament, the New Testament sees human nature as a psychophysical unity. Although Jesus and the apostles spoke Aramaic, their words have been handed down to us in New Testament Greek, which, depending on its frames of reference, uses any of several interchangeable terms for referring to persons. Whether we are admonished to love God with all our heart, soul, strength, and mind or to "present your bodies as a living sacrifice" the meaning is the same: commit your whole person to God.

The Greek word *psyche* parallels the Hebrew *nephesh* and is frequently translated as "soul." In many cases its meaning is clearly *not* that of an immaterial soul. When Joseph brought his father Jacob and 75 "souls" into Egypt in Acts 7:14 he did not leave their bodies behind in Canaan. The rich farmer dreams of harvests so great that he can say to his *psyche*, "Soul, you have ample goods laid up for many years; take your ease, eat, drink, and be merry." Biblical scholar Frank Stagg wonders aloud,

What kind of soul is it than can eat, drink, and be merry? A soul is a self, a person. In Rom. 2:9 every "human being" who does evil and suffers for it is a *psyche* and in Rom. 13:1 every "person" to be subjected to persons who govern is likewise a *psyche*. The whole [person] sins and the whole [person] is called to responsible citizenship. Paul, true to his Hebrew heritage, here thinks of man as a unity. . . . The Biblical teaching is not that one has a soul but that he is a soul.

Spirituality, as reflected in the Greek words translated "spirit" and "flesh" similarly has not to do with an invisible essence that is plugged into a bodily compartment, but with the whole person in relationship with God and other persons. Theologian Bruce Reichenbach suggests that to recapture this sense of spirituality we ought to drop the term *soul* from our religious vocabulary: "Such an approach, far from destroying faith in the spiritual aspect of man, will aid in clarifying precisely wherein the spiritual lies, that is, that it lies not in the possession of an entity, but in the style of life one leads insofar as it manifests a relation to God and to one's fellow man."

We also see the Hebrew-Christian understanding of psychophysical unity in the New Testament teaching concerning life after death. Oscar Cullmann begins his classic book, *Immortality of the Soul or Resurrection of the Dead?* by observing that if we were to ask ordinary Christians what they conceive to be the New Testament teaching concerning our fate after death, "with few exceptions we should get the answer: The immortality of the soul. Yet this widely accepted idea is one of the greatest misunderstandings of Christianity."

For Jesus, unlike Socrates, death was no friend. At the grave of his friend Lazarus, Jesus wept. Death mattered. It was, in the apostle Paul's words, "the great enemy." Death is real, and it is an enemy precisely because we do not have within our own natures a guaranteed immortality. At the end of our lives we do not, as Socrates assumed, "pass away"; rather, we die.

But there is hope, a hope rooted not in our nature but in God's love and faithfulness. Christians believe that God created and values human lives and that God will recreate them after death, giving us on that "great gettin-up morning" what, apart from divine love, we do not have—eternal life. The hope that Christians proclaim in the Apostle's Creed—"I believe in the resurrection of the body and the life everlasting"—is a hope grounded in God's initiative, not in our nature. To use a crude but modern analogy, after the plug is pulled on our computing machinery the divine programmer promises to recreate our software on a new, error-free piece of hardware. (Contrast this view with the pre-Christian idea of Seneca, who viewed himself as a "mixture of body and soul, of divine and human; my body I will leave where I found it, my soul I will restore to heaven.")

C. S. Lewis reminds us that if we have immortal souls then it must not have been the case that Christ was the first to defeat death, nor did he need to force open a door that until then had been locked. But Christians believe that it was and he did. Thus Lewis argued in his *Miracles* "that if the Psychical Researchers succeeded in proving 'survival' and showed that the Resurrection was an instance of it, they would not be supporting the Christian faith but refuting it." (In fact, the founders of parapsychology were mostly people who had lost their faith in God and were searching for another basis for believing in the meaning of life and the possibility of life after death.) Lewis embodies the Christian concept of resurrection in the first of his *Chronicles of Narnia*. When the wicked witch turns creatures into stone statues, they cease to exist as conscious beings. But the ultimate victory belongs to the great lion Aslan. After sacrificing his life, Aslan breaks the bonds of death and returns to give new life to the statues. "Bless me!" declares the giant Rumblebuffin, "I must have been asleep." So far as Rumblebuffin was concerned, he moved as instantly from death to paradise as a sound sleeper moves from the night to the morning.

Of the details of this resurrected body and life we need not worry. The points worth remembering are instead these: First, that our lives will be followed not by eternal extinction but by a renewal of life, with our individual identities intact, perhaps rather as a beautiful flower preserves the identity of the humble seed that precedes it. (From this, all Christians, whether they hold to an immortal soul or not, derive equal comfort when confronting death.)

Second, the New Testament image of a restored and perfected mind-body unit reinforces the other biblical images of human nature as a psychophysical unity. We must be wary of yoking biblical ideas to the details of any currently prevailing scientific theory. But it is noteworthy that this unified image is consistent with the emerging scientific image of humans as a mind-body unity. Fundamentally, both views assume that without our bodies we are nobodies, and that we had best therefore be good to our bodies. Rather than despising the body as that which "fills us with passions, and desires, and fears, and all sorts of fancies and foolishness," as Socrates declared, Christians regard the body as "the temple of the Holy Spirit." Indeed, we do not *have* bodies, we *are* bodies, bodies alive with minds.

For Further Reading

Reichenbach, B. *Is Man the Phoenix? A Study of Immortality.* Grand Rapids, Mich.: Eerdmans, 1978.
> For those wondering about particular verses that suggest a dualism of body and soul, or wishing to think more about what the Bible really says about human nature, Reichenbach's little book is must reading.

Cooper, J. "Dualism and the Biblical View of Human Beings," *Reformed Journal* (September, October 1982): 13–16, 16–18.
> Philosopher Cooper rebuts arguments against the dualism of body and soul.

Chapter 6

ON LIVING PEACEABLY WITH THE MYSTERIES OF FAITH

For my thoughts are not your thoughts
   neither are your ways my ways, says the Lord.
For as the heavens are higher than the earth,
   so are my ways higher than your ways
   and my thoughts than your thoughts.

ISAIAH 55:8–9, RSV

If the late, great developmental psychologist Jean Piaget is remembered for anything a century from now, it will likely be for his demonstrations that, as Isaiah might have put it, our thoughts are not a young child's thoughts. As a bird soars higher than a cat, so are an adult's thoughts higher than a child's.

By asking children questions and playing ingenious little games with them, Piaget discovered some remarkable and consistent errors in the way children of a given age interpret the world. If he would toss his beret over an appealing toy, a six-month-old infant would react as though the toy had ceased to exist. A one year old seems to understand that objects have permanence but cannot yet represent things with words. A three year old can abstract things in words but is unable to perceive things from another's perspective (such as your view of the television while the child is standing in front of it). Likewise, a five year old is becoming less egocentric but still has difficulty grasping analogies or reversing such mental operations as adding and then subtracting, while an eight year old can grasp simple analogies and perform mental operations but cannot reason with the logical powers of an adolescent. "When I was a child, I spoke as a child, I thought like a child, I reasoned like a child; when I became a man, I gave up childish ways."

From this insight that the child's mind is not simply that of a miniature adult we can derive two lessons. First, as parents, teachers, or Christian educators we may wish to be sensitive to children's limitations. Hard as it is for the adult to appreciate, preschoolers may be forming mostly misconceptions—which must later be reversed—of the meaning of Bible stories that adults love to teach them. Young primary school children may be incapable of grasping the analogy on which the object lesson of the children's sermon is based. When we try to pour gallon-sized concepts into pint-sized minds we should not be surprised when our children come home and tell us about "Gladly, the cross-eyed bear," or that God's name is "Hollywood," as in "Our Father which are in heaven, 'Hollywood' be thy name."

As children's minds develop so do their conceptions of God. They put away childish things such as their conceptions of a Santa Claus–like diety—which may not be exactly what they were taught but rather what they *thought* they were taught. Some revert to alternative simplistic images of God and the world. "We try to domesticate God," observes Madeline L'Engle, "to make his mighty actions comprehensible to our finite minds." Others will struggle with the competing truth–claims of various religions, with the problem of evil and the suffering of the innocent, with the clash between scientific findings and literal interpretations of biblical texts, and will reject their childish faith. Still others will come to recognize the immaturity of their childish world views *and* of adult views as well. Without absolute certainty about anything, they may then take the leap of faith.

Piaget's insights into children's difficulties in grasping our thoughts and ways also suggest a second valuable lesson. If God's thoughts and ways are higher than our own (as a jet flies higher than a bird), then *God is to us as we are to the preschooler, only more so.* Just as the preschooler cannot fathom adult logic, indeed is baffled by things that adults easily understand, so we can expect to be baffled by mysteries and paradoxes that are, perhaps, mere simplicities to God. Our position before God is rather like that of the occupant of a two-dimensional flatland trying to understand our three-dimensional world, or like our trying to conceptualize a world with four dimensions.

Try as we might, we can no more think our way through things that are to us ultimate riddles than a four year old can do calculus. If God is all-powerful *and* all-good, then why does evil exist? (The classic dilemma: either God cannot abolish evil or he will not: if he cannot, he is not all-powerful; if he will not, he is not all-good.) Or try this one: If God is the sovereign creator and sustainer of all history, then what room is there for human freedom? If there is even an ounce of human freedom, enabling history to be deflected this way or that at different forks in the road, then how can God be its sovereign Lord? If, on the other hand, God is ultimately in control of everything, even of our choices, then how can we humans be deemed responsible?

Such issues—called contradictions by nonbelievers and paradoxes by persons of faith—are indeed troubling, but less so once we realize that if God's thoughts and ways were like our own God wouldn't be God, or else we would be gods, too.

C. S. Lewis invited us to image a dog caught in a trap or a child with a thorn in a finger. To aid either we must ask them to trust what their limited intelligence cannot comprehend: that moving the paw farther back into the trap is the way to get it out, that hurting the finger more may be the way to stop its hurting. We can only hope that, based on nothing besides their confidence in us, the dog and the child will have faith. "Sometimes, because of their unbelief, we can do no mighty works," noted Lewis. Nevertheless, what is appropriate behavior for the dog and child is appropriate for us:

If human life is in fact ordered by a beneficent being whose knowledge of our real needs and of the way in which they can be satisfied infinitely exceeds our own, we must expect *a priori* that His operations will often appear to us far from beneficent and far from wise, and that it will be our highest prudence to give Him our confidence in spite of this.

Make no mistake about what we are suggesting. We are *not* saying give up the struggle, do not doubt, stop trying to turn childish beliefs into more mature ones. The Old Testament heroes of faith were people who dared admit their bafflement, who even dared argue with God. To immediately shrug off every difficult question by saying that we can't know God's thoughts is not so much intellectual humility as it is a cop-out. Some baffling issues may be neither inherent contradictions nor paradoxes, but simply unresolved puzzles.

But if, having pondered, searched, and struggled we remain baffled, we can relax. To our finite minds some philosophical puzzles seem impenetrable. At such points, science may actually be an aid to faith, both by reminding us of the immaturity of our cognition (on a divine scale) and by suggestng that irreconcilable concepts may, from our perspective, be an essential characteristic of nature. Light is a wave and light is a particle, the physicists tell us. "There are trivial truths and great truths," said the physicist Niels Bohr. "The opposite of a trivial truth is plainly false. The opposite of a great truth is also true."

The Danish philosopher-theologian, Søren Kierkegaard, noted that after grappling with the paradoxes and contradictions of faith we are left a " 'frightful act of decision.' The choice is between unbelief, which sees sheer madness in the affirmations of faith, and belief, which sees in that madness the divine wisdom." To live with the mysteries of faith requires that we do not demand of God that we be able to comprehend his being. We must in the last analysis accept that as the heavens are higher than the earth or as a mature adult's understanding is higher than a toddler's, so God's ways are higher than ours.

For Further Reading

Evans, C. S. *The Quest for Faith: Reason and Mystery as Pointers to God*, Downers Grove, Ill: InterVarsity Press, 1986.
　　　Philosopher Evans takes just 137 pages to explain why the Chris-

tian faith is reasonable and intellectually defensible, and to debunk many common skeptical arguments.

Fowler, J. *Stages of Faith: The Psychology of Human Development and the Quest for Meaning.* San Francisco: Harper & Row, 1981.
Fowler theorizes about how faith develops and how religious thinking changes across the life span.

Parks, S. *The Critical Years: The Young Adult Search for a Fdith to Live By.* San Francisco: Harper & Row, 1986.
Applies the "stages of faith" model to the critical college and young adult years.

Chapter 7

HOW MUCH CREDIT (AND BLAME) DO PARENTS DESERVE?

Train up a child in the way he should go,
   and when he is old he will not depart from it.
PROVERBS 22:6, RSV

Everyone believes it: by instruction, by discipline, and by example parents shape their children. To be convinced, most of us need look no further than our own families. We see ourselves reacting to situations much as our mother or father did. We hear their admonitions echoing in our minds. We relish their approval. We carry forward many of their values. And we see ourselves—or hope to see ourselves—not only reaching backward into our parents but forward into our children, chips off ourselves.

Countless research studies confirm the potency of parenting. The extremes of parenting provide the clearest evidence: the abused children who later become abusive, the unloved who become unloving. Orphanage-reared infants who are given minimal custodial care—ample food and a warm bed, but not much else—often become pathetic creatures, withdrawn, frightened, even speechless. By contrast, children who develop a positive self-image and a happy, self-reliant manner tend to have been reared by caring parents who are neither permissive nor autocratic, parents who have firm standards without depriving their children of a sense of control over their own lives.

In many ways we can see the parent in the child. By ten months of age, our babbling mirrors the sounds and intonations of our parents' language. In childhood, our dress, our play, and our ambitions look suspiciously like that of our same-sex parent. As adolescents, most of us still express the social, political, and religious views of our parents; the generation gap typically involves nothing more than differences in the strength with which we and our parents hold our shared values.

So we know both from experience and from the accumulating evidence the parental power that was understood by the writer of Proverbs. How one trains up a child affects how the child relates, talks, dresses, thinks, believes.

Our assumptions about the power of positive parenting lead us to credit parents for their children's achievements and blame them for their children's shortcomings. We may think how *we* would have handled that troubled child—surely with better results. The 1987 president of the Detroit city council even proposed that parents be punished for the criminal activities of their children.

Likewise, parents take personal pride in their children's successes and feel guilt over their failures. Parents accept congratulations for the child who is elected class president and feel shamed by the child who repeatedly is called to the principal's office. Parents second-guess themselves—where did they go wrong with him? How should they have handled her? It all makes perfect sense: if parents form children as a potter molds clay, then parents can indeed be praised for their children's virtues and blamed for their children's vices.

Given our readiness to praise or blame, to feel pride or shame, we had best also understand parental impotence. For the accumulating evidence further testifies to the ways in which children are shaped by forces over which parents have little control. One such force lies hidden within our genes, the architectural codes directing biochemical events that, down the line, determine our bodies and influence our behaviors. By selective breeding of animals, by comparing the similarity of genetically identical twins with that of fraternal twins, and by asking whether adopted children more closely resemble their biological or adoptive parents, psychologists are discovering how we are influenced by our heredity. Although disentangling the effects of genes and experience is no easy matter, it more and more seems that the genetic influence is considerable. The range of genetically influenced traits is impressive—from physical traits (such as handedness and obesity-proneness), to intelligence, to aggressiveness, to our vulnerability to depression and schizophrenia. In one study of 850 twin pairs, John Loehlin and Robert Nichols found that, compared to fraternal twins, identical twins were much more similar to one another in abilities, personality, and interests. It's true that identical twins often are dressed and treated more similarly than are fraternal twins, but Loehlin and Nichols found that twins whose parents treated them very similarly were *not* more alike than those who were treated less similarly. Even twins who are reared apart exhibit amazing similarities of tastes, personalities, and abilities. "In some domains it looks as though our identical twins reared apart are . . . just as similar as identical twins reared together," reports investigator Thomas Bouchard.

We must be careful not to oversimplify genetic effects. Our genes issue orders for our bodies, but our humanity also embodies nurturance provided or withheld, education given effectively or poorly, love sustained or withdrawn. Moreover, as every student of psychology knows, our personality reflects the interaction of our genes, past experience, and present situation. If a slow-witted, frail, uncoordinated boy experiences failure in the classroom, on the athletic field, and in his relations with girls, shall we say his low self-image is due to his genes or his environment? It's both, because his environment reacts to his genetically influenced traits.

Studies of adoptive families further restrain our belief in the unilateral power of parenting. The astonishing result of these studies is that the personalities of people who grow up together do *not* much resemble one another, whether they are biologically related or not. To be sure, adoption has some wonderful consequences: it transmits values and attitudes, and it provides a nurturant environment for children who might otherwise be hindered by neglect or abuse. Nevertheless, some dimensions of personality, such as temperamental reactivity, seem not to be greatly affected by normal variations in parenting. Developmental psychologist Sandra Scarr puts it more shockingly: "Our studies suggest that there's virtually no family environment effect on personality. . . . These data say that in any reasonable environment, people will become what they will become."

Although the evidence of parental power tempers Scarr's sweeping generalization, there are additional influences over which parents have little voluntary control. One such influence is the power of peers, which begins to supplant parental influence during adolescence. (The best predictor of whether a high school student smokes marijuana is simply how many of the student's friends smoke it.) Parents may therefore feel powerless to influence their teenager's tastes in music, dress, and recreation.

Lives also are formed by unpredictable events—by the jobs one happens to have held, the schools one happens to have attended, the illnesses or tragedies one happens to have suffered, even the chance encounters one happens to have experienced. Such life events deflect us down one road or another, and that can make all the difference. This, too, may explain why brothers and sisters who grow up in the same family are so often unlike one another, and why researchers who have followed lives through time are sometimes struck by the unpredictability of the life course. After following 166 lives from babyhood to age thirty, Jean Macfarlane was astonished to observe, "Many of our most mature and competent adults had severely troubled and confusing childhoods and adolescences." Often, the unhappy, troubled, rebellious adolescent became a stable, successful, happy adult.

In that fact we parents and parents-to-be can take comfort. It may, however, be discomforting to realize that having and raising children is a risky business; in procreation a man and woman shuffle their gene decks and deal a life-forming hand to their child-to-be, who thereafter is subjected to countless influences beyond their control. But perhaps we may also take comfort in knowing that we are therefore responsible not for our children's behavior, but for having given them our best. "Train up a child in the way he should go," and then love and accept what results.

When thinking about particular families we also do well to remember that the proverbial admonition is complemented by Jesus' admonition, "Judge not." Remembering that lives are formed by influences under parents' control *and* by influences beyond parents' control, let us be slow to credit parents for their children's achievements and slower still to blame them for their children's problems. Likewise, let us restrain our vanity when our children succeed and our feelings of guilt when they fail. As parents let us train up our children in the way they should go, and let us be slow to judge one another.

For Further Reading

Chess, S., Thomas, A., and Birch, H. G. *Your Child is a Person*: A *Psychological Approach to Parenthood without Guilt* (New York: Penguin, 1977).
A readable classic that recognizes that children come with particular temperamental traits.

PART 4 / **Sensation and Perception**

Chapter 8

THE MYSTERY OF THE ORDINARY

"As a rule," said Holmes, "the more bizarre a thing is the less mysterious it proves to be. It is your commonplace, featureless crimes which are really puzzling."

SIR ARTHUR CONAN DOYLE,
THE ADVENTURES OF SHERLOCK HOLMES, 1891–1892

At the core of the religious impulse is a sense of awe, an attitude of bewilderment, a feeling that reality is more amazing than everyday scientific reasoning can comprehend. Wonderstruck, we humbly acknowledge our limits and accept that which we can not explain.

For many religious people the ultimate threat of science is therefore that it will demystify life, destroying our sense of wonder and with it our readiness to believe in and worship an unseen reality. Once we regarded flashes of lightening and claps of thunder as supernatural magic. Now we understand the natural processes at work. Once we viewed certain mental disorders as demon possession. Now we are coming to discern genetic, biochemical, and stress-linked causes. Once we prayed that God would spare children from diphtheria. Now we vaccinate them. Understandably, some Christians have come to regard scientific naturalism as "the strongest intellectual enemy of the church."

We can also understand why such people therefore grasp at hints of the supernatural—at bizarre phenomena that science cannot explain. Browse your neighborhood religious bookstore and you will find books that describe happenings that defy natural explanation—

people reading minds or foretelling the future, levitating objects or influencing the roll of a die, discerning the contents of sealed envelopes or solving cases that have dumfounded detectives. Whether viewed as a divine gift or as demonic activity, such phenomena are said to refute a mechanistic world view that has no room for supernatural mysteries.

For several reasons, most research psychologists and professional magicians (who are wary of the exploitation of their arts in the name of psychic powers) are skeptical: (1) in the study of ESP and the paranormal there has been a distressing history of fraud and deception; (2) most people's beliefs in ESP are now understandable as a by-product of the efficient but occasionally misleading ways in which our minds process information (more on this in Chapter 14); (3) the accumulating evidence regarding the brain-mind connection more and more weighs against the theory that the human mind can function or travel separately from the brain; and, most importantly, (4) there has never been demonstrated a reproducible ESP phenomenon, nor any individual who could defy chance when carefully retested. After one hundred years of research, and after hundreds of failed attempts to claim a $10,000 prize that has for two decades been offered to the first person who can demonstrate "*any* paranormal ability," many parapsychologists concede that what they need to give their field credibility is a single reproducible phenomenon and a theory to explain it.

We Christians can side with the scientific skeptics on the ESP issue. We can heed not only the repeated biblical warnings against being misled by self-professed psychics who practice "divination" or "magic spells and charms," but also the scientific spirit of Deuteronomy: "If a prophet speaks in the name of the Lord and what he says does not come true, than it is not the Lord's message." We believe that humans are finite creatures of the one who declares "I am God, and there is none like me." We are aware of how cult leaders have seduced people with pseudopsychic tricks. And we affirm that God alone is omniscient (thus able to read minds and know the future), omnipresent (thus able to be in two places at once), and omipotent (thus capable of altering—or better yet—creating nature with divine power). In the biblical view, humans, loved by God, have dignity, but not deity.

If our sense of mystery is not to be found in the realm of the pseudosciences and the occult, then where? Having cleared the decks of false mysteries, where shall we find the genuine mysteries of life? We can take our clue from Sherlock Holmes who was fond of telling people,

It is a mistake to confound strangeness with mystery. The most common-place crime is often the most mysterious . . . Life is infinitely stranger than anything which the mind of man could invent. We would not dare to conceive the things which are really mere commonplaces of existence.

The more scientists learn about sensation, the more convinced they are that what is truly extraordinary is not extrasensory perception, claims for which inevitably dissolve upon investigation, but rather our very ordinary moment-to-moment sensory experiences of organizing formless neural impulses into colorful sights and meaningful sounds. As you read this sentence, particles of light energy are being absorbed by the receptor cells of your eyes, converted into neural signals that activate neighboring cells, which process the information for a third layer of cells, which converge to form a nerve tract that transmits a million electrochemical messages per moment up to the brain. There, step by step, the scene you are viewing is reassembled into its component features and finally—in some as yet mysterious way—composed into a consciously perceived image, which is instantly compared with previously stored images and recognized as words you know. The whole process is rather like taking a house apart, splinter by splinter, transporting it to a different location, and then, through the work of millions of specialized workers, putting it back together. All of this transpires in a fraction of a second. Moreover, it is continuously transpiring in motion, in three dimensions, and in color. Ten years of research on computer vision has not yet begun to duplicate this very ordinary, taken-for-granted part of our current experience.

Further, unlike virtually all computers, which process information one step at a time, the human brain carries out countless other operations simultaneously, enabling us all at once to sense the environment, use common sense, converse, experience emotion, and consciously reflect on the meaning of our existence or even to wonder about our brain activity while wondering. The deeper one explores these very ordinary things of life, the more one empathizes with Job: "I have uttered what I did not understand, things too wonderful for me."

To be sure, sometimes we use the word *mystery* not in its deep sense as when the mind seeks to fathom its brain, but rather to refer to unsolved scientific puzzles. When wonder is based merely on ignorance, it will fade in the growing light of understanding. Science is a puzzle-solving activity. Among the still unsolved puzzles of psychology are questions such as, Why do we dream? Why do some of us become heterosexual, others homosexual? How does the brain store memories? The scientific detectives are at work on these "mysteries," and they may eventually offer us convincing solutions. Already, new ideas are emerging.

Often, however, the process of answering one question exposes more and sometimes deeper questions. A new understanding may lead to a new, more impenetrable sense of wonder regarding phenomena that seem further than ever from explanation or that now seem more beautifully intricate than previously imagined. Not long ago scientists wondered how individual nerve cells communicated with one another. The answer—through chemical messengers called neurotransmitters—raised new questions: How many neurotransmitters exist? What are the functions of each? Do abnormalities in neurotransmitter functioning predispose disorders such as schizophrenia and depression? If so, how might such problems be remedied? And how, from the electrochemical activity of the brain, do experienced emotions and thoughts arise: *how does a material brain give rise to consciousness?* Deeper and deeper go the questions, the deepest one of all being the impenetrable mystery behind the origin of the universe: *why is there something and not nothing?* (If a miracle is something that cannot be explained in terms of something else, then the existence of the universe is a miracle that dwarfs any other our minds can conceive.)

Human consciousness has long been a thing of wonder. More recently, wonder has also grown regarding the things our minds do subconsciously, automatically, out of sight. Our minds detect and process information without our awareness. They automatically organize our perceptions and interpretations. They respond (via the right hemisphere) intelligently in ways that we can explain only if our left hemisphere is informed what is going on. They effortlessly encode incoming information about the place, timing, and frequency of events we experience, about word meanings, about unattended stimuli. They ponder problems we are stumped with, and they occasionally spew forth a spontaneous creative insight. With the aid of hypnosis, they may even, on orders, eliminate warts on one side of the body but not on the other. There is, we now know, more to our minds than we are aware. And how fortunate that it should be so. For the more that routine functions (including well-learned activities such as walking, biking, or gymnastics) are delegated to control systems outside of awareness the more our consciousness is freed to function like an executive—by focusing on the most important problems at hand. Our brains operate rather like General Motors, with a few important matters decided by the chair of the board, and everything else, thankfully, handled automatically, effortlessly, and usually competently by amazingly intricate mechanisms.

The more they explore, the more language reseachers, too, have been awestruck by an amazing phenomenon: the ease with which children acquire language. Before children can add two plus two they are creating their own grammatically intelligible sentences and comprehending the even more complex sentences spoken to them. Most parents cannot state the intricate rules of grammar, and they certainly are not giving their children much formal training in grammar. Yet before being able to tie their shoes, preschoolers are soaking up the complexities of language by learning several new words a day and the rules for how to combine them. They do so with a facility that puts to shame many college students who struggle to learn a new language with correct accents and many computer scientists who are struggling to simulate natural language on computers. Moreover, they, and we, do so with minimal comprehension of how we do it—how we, when speaking, monitor our muscles, order our syntax, watch out for semantic catastrophes risked by the slightest change in word order, continuously adjust our tone of voice, facial expression, and gestures, and manage to say something meaningful when it would be so easy to speak gibberish.

Our womb-to-tomb individual development is equally remarkable. What is more ordinary than humans reproducing themselves, and what is more wonder-full? Consider the incredible good fortune that brought each one of us into existence. The process began as a mature egg was released by the ovary and as some 300 million sperm began their upstream race towards it. Against all odds, you— or more exactly the very sperm cell that together with the very egg it would take to make you—won this 1 in 300 million lottery (actually one in billions, considering your conception had to occur from that particular sexual union). What is more, a chain of equally improbable events, beginning with the conception of your parents and their discovery of one another had to have extended backwards in time for the possibility of your moment to have arrived. Indeed, when one considers the improbable sequence of innumerable events that led to your conception, from the birth of the universe onward, one cannot escape the conclusion that your birth and your death anchor the two ends of a continuum of probabilities. What is more *im*probable than that you, rather than one of your infinite alternatives, should exist? What is more certain than that you will not live on earth endlessly?

Most beginnings of life fail to survive the first week of existence. But, again, for you good fortune prevailed. Your one cell became two, which became four; and then by the end of your first week an even more astonishing thing happened—brain cells began forming and within weeks were multiplying at a rate of about one-quarter million per *minute*. Scientist-physician Lewis Thomas explains the wonder of that single cell, which has as its descendants all the cells of the human brain.

The mere existence of that cell should be one of the great astonishments of the earth. People ought to be walking around all day, all through their waking hours, calling to each other in endless wonderment, talking of nothing except that cell. . . .

If you like being surprised, there's the source. One cell is switched on to become the whole trillion-cell, massive apparatus for thinking and imagining and, for that matter, being surprised. All the information needed for learning to read and write, playing the piano, arguing before senatorial subcommittees, walking across a street through traffic, or the marvelous human act of putting out one hand and leaning against a tree, is contained in that first cell. All of grammar, all syntax, all arithmetic, all music. . . .

No one has the ghost of an idea how this works, and nothing else in life can ever be so puzzling. If anyone does succeed in explaining it, within my lifetime, I will charter a skywriting airplane, maybe a whole fleet of them, and send them aloft to write one great exclamation point after another, around the whole sky, until all my money runs out.

Human life—so ordinary, so familiar, so natural, and yet so extraordinary. Looking for mystery in things bizarre, we feel cheated when later we learn that a hoax or a simple process explains it away. All the while we miss the awesome events occurring before, or even within, our very eyes. The extraordinary within the ordinary.

So it was on that Christmas morning two millennia ago. The most extraordinary event of history—the Lord of the universe coming to the spaceship earth in human form—occurred in so ordinary a way as hardly to be noticed. On a mundane winter day at an undistinguished inn in an average little town the extraordinary one was born of an ordinary peasant woman. Like our human kin at Bethlehem and Nazareth long ago, we, too, are often blind to the mystery within things ordinary. We look for wonders and for the unseen reality—the hand of God—in things extraordinary, when more often his presence is to be found in the unheralded, familiar, everyday events of which life is woven.

For Further Reading

Brand, P., and P. Yancey. *Fearfully and Wonderfully Made.* Grand Rapids, Mich.: Zondervan, 1980.
  A wonderfully made and wonder-filled book about the human organism that expands on the theme of this chapter. If you like this, you will also enjoy their 1984 follow-up book, *In His Image.*

Thomas, L. *The Lives of a Cell; The Medusa and the Snail; Late Night Thoughts on Listening to Mahler's Ninth Symphony.* New York: Viking Press, 1974, 1979, 1983.
  Physician-scientist Thomas's collections of short essays describe scientific wonders with beautiful awestruck prose.

Chapter 9

ON DISCERNING SENSE FROM NONSENSE

The way must be prepared for your moral assaults by darkening his intellect.

SCREWTAPE TO WORMWOOD,
IN C. S. LEWIS'S *SCREWTAPE LETTERS*, 1942

In C. S. Lewis's tongue-in-cheek book of advice to a junior devil, Screwtape counsels his student to corrupt by diverting attention: "Success depends on confusing. . . . The game is to have them all running about with fire extinguishers whenever there is a flood."

Wormwood, the junior devil, still follows Screwtape's advice. One way he does so is by seducing people to fear a comic strip–like evil character who sneaks around doing devilish little things, such as taking possession of someone's body (until being exorcized), communicating through Ouija boards, or inserting backward messages on to rock records.

To some people, such claims are as likely as a devil who wears red tights and carries a pitchfork. To others, they represent the very mechanisms by which evil powers operate. Consider the supposed phenomenon of "backmasking"—satanic messages that are allegedly recorded backward into rock music by diabolical rock groups or by Satan himself, and which are said to unconsciously corrupt listeners. Throughout the United States and Canada, concerned ministers have warned parents and their children about the existence and corrupting power of these backward messages. In several American states and in the Canadian parliament, political lobbying has been undertaken to contain the damage. In the state of Arkansas, a bill was passed in 1983 requiring certain records and tapes to be sold with a message affixed: "Warning: This record contains backward masking which may be perceptible at a subliminal level when the record is played forward."

Does an evil power in fact corrupt people in this way? Or does it rather mislead people to *think* that it does, thereby diverting their attention from genuine evil? Like so many issues over which we fruitlessly argue and worry, this one is amenable to dispassionate research. A look at how two investigators approached this issue serves to ilustrate psychology's contribution to human understanding—its process of inquiry. Fortunately, the Creator, knowing that not even a heart of gold would substitute for a head of feathers, saw fit to put brains in our heads so that we might use them in disciplined efforts to winnow truth from error. The unmasking of the popular backmasking claim illustrates the difference between the speculations of pop psychology and the more modest but hard-won achievements of psychological science.

Having been asked by a local radio station to comment on a backmasking seminar and a subsequent record-smashing rally in their city of Lethbridge Alberta, research psychologists John Vokey and Don Read decided to investigate. Vokey and Read were well aware of claims of subliminal advertising (*subliminal* means below the threshold for conscious awareness). Movie audiences are rumored to have been unknowingly influenced by imperceptible messages to "eat popcorn." The word *sex* is said to have been subtly imbedded in advertising copy and even in products such as Ritz crackers—all in an effort to manipulate us.

Vokey and Read were also aware of recent experiments indicating that people can be influenced by exposure to stimuli that are not consciously perceived. Imagine that as a subject in one of these experiments you are asked which of two words, *penny* or *rabbit* is closest to a word that has been flashed on a screen. The flash is so brief that, consciously, you cannot even tell whether *any* word has appeared amidst the flash. Remarkably, however, your guess is more likely to be right than wrong. If so, you processed information from the flash with no awareness of having done so.

Because such results are a far cry from situations in which sub-liminal persuasion supposedly occurs—in which people's attention is already captured by other stimuli (a movie, a picture, a song)—psychologists remain unconvinced by claims of subliminal persua-sion. Nevertheless, most psychologists now concur that many—perhaps most—mental processes occur outside of consciousness awareness. So rather than close their minds to the possibility of an unconscious backmasking effect, Vokey and Read undertook an in-genious series of experiments.

As psychologists, they were in no position to answer the question, have backward messages in fact been inserted into rock albums? But they did have the tools to explore a related question: can people derive meaning from or be influenced by backward messages? If answered in the negative, this question would make the existence of such messages irrelevant.

They began by recording some simple sentences (such as from Lewis Carroll's "Jabberwocky" and from the Twenty-third Psalm) and then rerecorded them in the backward direction. Because the pauses, pitches, and changes in emphasis were retained, the result was a speechlike recording that sounded rather like a novel foreign language. College students asked to identify the sex of the speaker could nearly always do so and could slightly surpass chance in guessing whether the forward version of the message was English, French, or German.

Could they also decode meaning? Asked simply to estimate how many words were in a spoken sentence, the listeners were influ-enced by the number of syllabic breaks rather than by the number of words that were in the message. Moreover, when asked whether a given word had been played backwards in a sentence, their 56 percent rate of correct responding barely surpassed the chance level of 50 percent correct. These initial tests suggested that virtually none of the forward meaning of a backward message can be con-sciously perceived.

Might it instead be *un*consciously perceived? Without being asked directly about the meaning of what they had heard, listeners were asked merely to say whether it had been a declarative statement or a question. Their guessing accuracy was a mere 52 percent—statistically indistinguishable from the chance level of 50 percent. Presented with pairs of backward sentences and asked whether they had a similar or different meaning, the listeners' 45 percent correct guessing was slightly below the 50 percent chance level. A similar result was obtained when listeners were asked whether they had heard a meaningful or a nonsensical sentence.

Recognizing that any one research method may give incomplete or misleading information, Vokey and Read persisted by searching for the backward masking phenomenon from every conceivable direction. They gave people backward recordings of simple messages such as "Jesus loves me, this I know" and solicited their hunches as to which of five categories the message belonged: nursery rhyme, Christian, satanic, pornographic, or advertising. Again, there was no evidence that any meaning had been gained subconsciously—the guessing rate of 19.4 percent was virtually identical to the 20 percent chance rate. So far as the researchers could discern, then, no meaning was gleaned from backward messages.

Might it nevertheless be possible that backward messages have a subtle influence? Vokey and Read did not want to experiment with messages advocating promiscuity or illicit drug use. So they instead capitalized upon a well-known vehicle for studying unconscious influence. When people hear sentences such as "Climbing a mountain is a remarkable *feat*" they are more likely later to spell the italicized homophone word as *feat* (rather than *feet*) than are people who haven't previously heard the word in a sentence. The effect occurs even when people are unaware of having previously heard the word being read. However, Vokey and Read found that the effect disappears totally when the homophones are embedded in backward rather than forward messages. The researchers' conclusion: there is no evidence that people derive meaning from or are influenced by backward messages.

Why, then, are those who are concerned about backmasking able to discern diabolical messages when rock music is played backwards? The answer lies in a phenomenon that is well known both to students of perception and to experienced cloud-watchers: given a *perceptual set* we tend to see or hear in an ambiguous stimulus what we expect to see or hear. The perception of the word *sex* on a Ritz cracker likely says more about the perceiver than the cracker.

To demonstrate, Vokey and Read devoted some (one suspects hilarious) hours to listening to their two favorite backward recordings—of "Jabberwocky" and the Twenty-third Psalm. Much to their delight, they were able to invent some phrases that could be interpreted as consistent with the sound sequences, phrases such as, "Saw a girl with a weasel in her mouth," and even "I saw Satan." They then gave listeners their phrases, one at a time, and asked them to listen for them in the two recordings. Usually the listeners were able to hear the phrases they were listening for (but only in the passage that contained a sequence of noises that was consistent with the suggestion). Had the listeners heard any of the phrases in any of their previous listenings before receiving the suggestion? Nearly always, they had not. So it is not surprising that in public demonstrations of backward messages, preachers do not play their backward tapes and ask their audiences what they heard; rather, they first tell people what to hear and then play and replay the segment where the expected words may be "heard."

Vokey and Read's findings became front-page news in the *Arkansas Democrat* as the backward masking bill lay on the governor's desk awaiting his signature. Subsequently, the bill was returned to the Arkansas senate for review and was promptly defeated. Perhaps the legislators awoke to realize that the real influence of rock music lies not in its backward but its forward messages, from which Wormwood had diverted their attention.

For Further Reading

Vokey, J. R. and J. D. Read. "Subliminal Messages: Between the Devil and the Media," *American Psychologist* 40 (1985):1231–39.
A full report on the controversy and the research described in this chapter.

Chapter 10

THROUGH THE EYES OF FAITH

The eyes of those who see
 will not be closed,
and the ears of those who hear
 will hearken.

One thing I know, that though
I was blind, now I see.

JOHN 9:25, RSV

W hy is it that two people hearing the same gospel, reading the same Scriptures, apprehending the same universe, can perceive things so differently? Some who observed Jesus' words and deeds perceived him as crazy: "He has a demon, and he is mad." Others saw him differently: "These are not the sayings of one who has a demon. Can a demon open the eyes of the blind?" What was foolishness to some was to others the power of God.

We can better understand how this might be if we view religious experiences as in some ways like other perceptual experiences. Perceptions arise from the interaction of *stimulus* and *perceiver*. A stimulus is detected and encoded by a person's senses, and this sensory information is then organized and interpreted by the brain. Out of this interaction between stimulus and perceiver, between what's out there and what goes on in our heads, emerges our perceptions.

Those who study human perception and its implications for scientific and religious knowledge have sometimes gravitated to one of two extremes. The subjectivist extreme discounts the importance of the objective stimulus. It sees our perceptions as arbitrary mental constructions, as meanings that we impose willy-nilly on the world out there. If that alone were true, we could not drive a car, nor could we hope to walk out of the room without walking into a wall. Our skill at responding to our physical environment assures us that there *is* an objective world out there and that we process its information with remarkable accuracy and efficiency.

The objectivist or naive realist extreme assumes that our experience mirrors reality. As we perceive it, so it is. By this view, truth can be achieved simply by checking our scientific theories against the perceived facts of nature and our religious doctrines against the perceived facts of Scripture. How you and I perceive something does indeed depend on the stimulus; our perceptions—and the theories and doctrines built upon them—are not arbitrary. But the extreme of this view discounts the importance of the state of the perceiver. Perceptions also depend on where our *attention* is drawn, on our prior *experience*, and on our *expectations*.

Because our consciousness has a limited capacity, we at any moment selectively *attend* to but a small corner of our experience. At a party, we attend to but one voice at a time, ignoring all the others. If you observe two people talking, only one of whom you can look in the face, the one whom you see will tend to capture your attention and to seem the more dominant. While reading this sentence, you attend to the words and, until you near its end, ignore the nose that intrudes in your line of vision. Attention constrains perception.

Out of our *experience* in the world, from infancy onward, we form schemas—ways of organizing and interpreting reality. Lacking a schema to interpret the black blotches of Figure 3, you probably make no sense of it. As you continue to stare at the figure your mind struggles to make sense out of the apparent chaos. With patience you eventually impose order, by seeing a dalmation dog on the right sniffing the ground at the center. Note that once your mind has this dog schema it controls your perception—so much so that it becomes virtually impossible *not* to perceive the dog. As this demonstrates, there is more to perception than meets the eye.

Finally, our assumptions and *expectations* may give us a perceptual set, a predisposition to interpret an ambiguous stimulus one way rather than another. In one famous experiment, illustrated in Figure 4, those shown the series of images in the top frame per-

Figure 3.

ceived the ambiguous figure at right as a face. Those shown in the lower series of images perceived the same image as a female figure. Same stimulus, differing perceptions.

In another experiment, 75 percent of those who were shown a slightly blurred picture could identify it. Others, having first seen a badly blurred picture, could identify the slightly blurred one only 25 percent of the time. Why do you suppose the second group had so much more trouble perceiving the true image?

The answer seems to be that once preliminary hunches were formed, based on the badly blurred picture, they interfered with accurate perceptions. Having formed a wrong idea about reality, people have more difficulty seeing the truth. As C. S. Lewis generalized, "What we learn from experience depends on the kind of

Figure 4.

philosophy we bring to experience." Whether perceiving visual stimuli, doing science, or reading Scripture, our expectations influence how we see things. To believe is to see.

For all these reasons, religious perceptions depend on the state of the perceiver as well as on external reality. Depending on one's perceptual set, a thought that pops into the mind while meditating may be perceived as a random cognition or as the still small voice of God. Moses perceived his burning bush and mountaintop experiences through the eyes of faith and thus assigned them a profound religious significance that would have been meaningless to someone lacking his perceptual set.

Imagine yourself looking with a friend at a clear night sky. Your friend points overhead and says, "Do you see the Little Bear?" Looking at the very same stars you cannot perceive what your friend so clearly sees. Why? Because your friend, having taken the trouble to study star patterns, has eyes to see what you are not ready to notice. Similarly, people may see the heavens, which declare the glory of God, yet not see that the heavens are declaring God's glory. As Pascal noted in his *Pensees*, only the heart that already has faith will see the heavens in that way. The point has been recognized even by religious skeptics, such as philosopher Paul Kurtz: "I have wondered at times: Is it I who lacks religious sense, and is this due to a defect of character? The tone-deaf are unable to fully appreciate the intensity of music, and the color-blind live in a world denuded of brightness and hue."

To have a religious experience is thus to assign to sensory experience spiritual significance. It is to interpret phenomena with an awareness of the presence of God. Those who have a schema for interpreting life in this way are like those who have the schema for perceiving the dalmation: they have difficulty viewing things any other way, yet sometimes find it hard to get others to see reality as they do.

With the schema of faith, a whole set of perceptions forcefully takes hold of one's consciousness. Jesus is perceived not as a psychotic but as an incarnation of God. The universe is seen not as a meaningless material reality, but as God's creative handiwork— the ultimate miracle that makes little sense without a Creator. Life itself takes on purpose in a world where humans are viewed as called to recognize their limits and their value to their Creator, to assume their responsibility for the earth and for one another's welfare, and to serve and enjoy God forever.

Lord, open our eyes that we may see.

PART 5

States of Awareness

Chapter 11

THE DAY OF REST

It is in silence, and not in commotion, in solitude and not in crowds,
that God best likes to reveal Himself most intimately to men.
THOMAS MERTON,
THE SILENT LIFE, 1957

Humans are by nature social animals. What is true of
nature in general—that no being exists on its own resources—is
plainly true of humans as well. All people, however independent
they may think themselves to be, are "hopelessly indebted to everyone and everything else." Babies, before and after birth, live on
their mothers and fathers. Beggars are dependent on the charity of
the rich. The rich in turn are dependent on the services of those
who make them rich and bring food to their table. Recognizing
our interdependence we value togetherness, communal fellowship,
group activities, team sports, close relationships, self-disclosure, social support. To be withdrawn, a hermit, a loner, is an aberration.

In the psychology of the 1980s one does, however, see a complementary awareness emerging. Too much social stimulation can
be stressful. Crowding, noise, sensory overload, loss of privacy, an-
onymity, social anxiety, fear of victimization, extreme competi-
tiveness, and impatience all take a toll on human well-being. What
is more, times of social solitude can heal and renew.

Such is the conclusion of an impressive series of studies con-
ducted by University of British Columbia researcher Peter Suedfeld
and his colleagues. Suedfeld knew from earlier studies of sensory
restriction that being alone in a monotonous environment
heightens a person's sensitivity to any external or internal stimuli.

So he offered hundreds of people an opportunity to tune more deeply into themselves through a twenty-four-hour experience of Restricted Environmental Stimulation Therapy—a literal day of REST. For a day, the person does nothing but lie quietly on a comfortable bed in the isolation of a dark and soundproofed room. Food, water, and a chemical toilet are available to service the body, and communication is possible over an intercom system through which brief persuasive messages may also be transmitted to the person.

The day of REST has been notably successful in assisting people who wish to increase their self-control—to gain or lose weight, reduce alcohol intake, improve speech fluency, reduce hypertension, overcome irrational fears, boost self-confidence, or stop smoking. People report that the experience is a pleasant and stress-free way of reducing external stimulation to a point where still, small internal voices can be heard.

The healing power of a period of aloneness can also be found in the lives of people who, by choice or necessity, have experienced periods of solitude. To be shipwrecked, placed in solitary confinement, or to be a solitary voyager can be traumatic if one feels threatened, helpless, or malnourished. But often there is a positive side to such experiences. The lone explorer or sailor may have a deep mystical experience—a new relationship with God, a feeling of oneness with the ocean or the universe, a life-changing new insight into his or her personality. If prisoners in solitary confinement are otherwise assured of humane care, free of privation and torture, they often alleviate the boredom by studying, thinking, solving personal problems, even planning their own rehabilitation.

Scores of cultures on the American, African, Asian, and Australian continents incorporate a period of solitude into the life history of every individual, or at least of every male. The boy entering manhood leaves his community to wander alone in the desert, mountains, forest, or prairie. During this time he searches his soul, dreams a vision, communicates with the gods, or experiences the oneness of nature. Through the experience the boy grows beyond his previous self to a new level of consciousness. Backpacking and Outward Bound adventuring (a month-long program of outdoor physical activities) can offer similar experiences. During the "Solo" component of Outward Bound and of many camping programs, the participants sometimes have mystical experiences that would do credit to any meditator. One recent study of 361 graduates of the twenty-six-day Australian Outward Bound program found that the experience produced enduring improvements in self-concept.

Traditional folk therapies for psychological disorders often have isolated the disturbed person for a time of aloneness. Many of today's institutions for mentally disturbed juvenile or adult offenders use "time out" rooms in which an agitated person experiences solitude. In Japan, the widely practiced "quiet therapies" combine solitude with traditions inherited from Zen Buddhism. The depressed or anxious person may commence therapy with a week of bed rest and meditation, after which activities are gradually reintroduced.

Suedfeld notes that many autobiographies and biographies confirm the creative power of solitude. Philosophers, scientists, and artists have experienced novel ideas while isolated. Freed from distraction and social demands, unusual things sometimes happen— vivid fantasies and memories, relaxed emotions, beautiful sensory experiences, deep insights.

The foremost examples are the great religious visions that have followed times of solitude and contemplation. Jesus, who began his ministry after forty days alone and who lived a rhythm of retreat and engagement, provides the most noteworthy example. Thousands of other religious visionaries, including Moses, Mohammed, Buddha, and countless mystics, monks, hermits, and prophets have found inspiration in times of contemplative silence. The Christian discipline of a daily quiet time affirms the value of restricted stimulation, not as an otherworldly end, but as spiritual recharging for living in this world. In times of silence and solitude, God reveals himself.

Finally, living as we are in a time when the hustle and bustle of working, shopping, and entertainment has become a seven-day-a-week affair we note an irony: our European and American cultures are turning away from the day of rest at the very time that researchers are affirming the healing and renewing power of a day of REST. And should such renewing power surprise us? We are, after all, made in the image of the one who on the seventh day "finished his work which he had done, and . . . rested. . . . So God blessed the seventh day and hallowed it, because on it God rested from all his work which he had done in creation."

For Further Reading

Suedfeld, P. *Restricted Environmental Stimulation: Research and Clinical Applications.* New York: Wiley, 1980.
    Suedfeld presents the whole story of his basic and applied research and of the history of human uses of REST.

Chapter 12

ARE WE DETERMINED OR FREE?

"The ideal reasoner," [Sherlock Holmes] remarked, "would when he has once been shown a single fact in all its bearings, deduce from it not only all the chain of events which led up to it, but also all the results which would follow from it."

SIR ARTHUR CONAN DOYLE,
THE ADVENTURES OF SHERLOCK HOLMES, 1891–1892

When Christians discuss psychology no issue arises more often than this: is human behavior determined? Imagine a question posed to two completely identical persons, two persons who are in every way—heredity, past experience, current brain states—replicas of one another. If we now confront them with identical choices in an identical manner (coffee or tea?) will each necessarily respond the same? Or could they act differently?

Note that the answer to this mind-teaser is either yes or no. As William James said, "The issue . . . is a perfectly sharp one, which no eulogistic terminology can smear over or wipe out. The truth *must* lie with one side or the other, and its lying with one side makes the other false."

An answer of "yes—there is nothing to differentiate them because all possible influences are identical" assumes determinism: behavior is lawfully related to causal influences. Although the people each made conscious choice, it was not in their power to have chosen otherwise.

Many psychologists—most we suspect—would answer yes. But others would answer no. Some of these reject determinism because they assume an element of *indeterminism*—of inherent unpredictability. Much as elementary particles behave with apparent randomness, so might human behavior exhibit a lack of orderly causation. Others answer no because they believe that just as God is the ultimate source of natural events, so are people, to some extent, the ultimate cause of their own actions. To be sure, we all are influenced by various biological and psychological factors; still, say the proponents of *agent causation*, these factors don't totally determine our behavior. When all is said and done, you and I can tip the scales this way or that, toward coffee or tea, toward moral or immoral action.

Partial Determinism as a Working Assumption

We psychologists agree that our work requires some regularity. We need not assume that behavior is completely determined to look for what orderly causes or predictors there are. To search out the factors that do, in fact, influence behavior, we don't need to make a philosophical assumption of absolute determinism; we only need to assume, as a tentative working hypothesis, enough determination to provide a detectable regularity. Whether we study perception, intelligence, depression, or aggression, there are three types of possible determinants that we might investigate: biological influences, past experience, and the current situation. (As we have already noted, some would add agent causation as an independent fourth causal factor.)

Biological Factors   *Current Situation*

PERSON

Past Experience

Evidence is accumulating regarding each of these three types of determinants. We now understand better than ever how our behavior is predisposed by *biological factors*. Our genes operate by directing biochemical events that, through the workings of our individual nervous systems, influence our thoughts, emotions, and behaviors. As we noted in Chapter 4, every advance in biopsychology has served to tighten the links between our genes, our brains, and our behaviors. If a newborn's brain is malnourished during the period when neural interconnections should be rapidly growing, the result may therefore be permanent impairments of memory, language, and perception.

Evidence also continues to accumulate concerning the effects of *past experience*. Harry and Margaret Harlow's classic studies revealed that early maternal deprivation could produce devastating effects on the later social behaviors of monkeys—a finding that is tragically paralleled by studies of neglected or abused children. More recent evidence also indicates that development is a lifelong process. Although later experiences never fully erase the voices of the past, they may reshape behavior in new directions—sometimes transforming an unhappy, confused, and rebellious young adolescent into a thriving adult.

We need look no further than at our own behavior in different situations—at a party, at worship, in class, with parents, with friends—to recognize that the *current situation* is a similarly powerful determinant. In *The Mountain People*, anthropologist Colin Turnbull records how a whole culture changed in response to a changed situation. The Ik people of northeastern Uganda were at one time a peaceful, cooperative society. But when the Ugandan government made the Ik's tribal grounds into a national park and moved them to a new mountainside area, they became a "passionless, feelingless association of individuals." They would laugh when a young child grabbed a hot coal, beat their elders, and, as starvation set in, hoard food and abandon old people to die. Theologian Langdon Gilkey's fascinating description of the *Shantung Compound*, a World War II internment camp into which the Japanese military herded foreigners residing in China, portrays a similar cultural degeneration. Being reduced to the bare essentials triggered a pervasive self-centeredness among virtually all its inhabitants—missionaries, doctors, lawyers, professors, businespeople, junkies, and prostitutes. Our behavior is strongly influenced by the context in which it occurs.

Absolute Determinism

As a working hypothesis, it has proven fruitful to assume an underlying regularity to behavior. Within the complexity of human nature, there is discernible order. When combined and interacting, biological, historical, and situational factors powerfully influence behavior.

How powerfully? Two of the most famous names in twentieth-century psychology, Sigmund Freud and B. F. Skinner, adopted determinism not only as a working assumption that gives meaning to a science of behavior and mental processes, but also as a philosophy. In important ways, Freud and Skinner differed: Freud used subjective methods of inquiry and emphasized the controlling power of unconscious residues of early life experience; Skinner experimentally studied the precise ways in which behavior is shaped by punishers and, especially, by rewards. But on this much they agreed: freedom is an illusion. To our mind-teaser question—will behavior be the same if all the determinants are the same?—both would answer, "Absolutely yes."

To reemphasize, as research psychologists, we need not assume absolute determinism. We need only assume what is now beyond question—that there is considerable order and regularity to human behavior. For our research it matters little whether this order is rooted in an absolute determinism or whether random indeterminacies or agent causation means that we can only hope to describe behavior in terms of probabilities. In either case, the enormous complexity of human nature will always limit us to statistical generalizations. Compared to predicting people, predicting the weather is easy.

Nevertheless, let's imagine just for a few moments that our behavior is absolutely determined and therefore, in principle, predictable to an all-knowing and all-wise scientist—the "ideal reasoner" whom Sherlock Holmes had in mind. What then? Are the implications as terrible as most people suppose? (Note: we are not advocating absolute determinism, because we don't know what the ultimate truth is; we are simply attempting to clear away some misunderstandings about determinism.)

One approach to this question is to consider the opposite—a world with *no* determinism and therefore, possibly, with utter unpredictability. Is it not in such a world, rather than the absolutely determined world of Freud and Skinner, that people would adopt a fatalistic whatever-will-be-will-be attitude? If people's actions tomorrow were *not* influenced by the circumstances created today, then nothing we do today could make a difference. To act responsibly, we must have some idea of the effects of our actions. Indeed, it is in a world in which our actions *do* have predictable effects that hope reigns eternal. If "a man reaps what he sows," then we have a responsibility for the future. In a deterministic world the stream of causation runs from the past to the future through our choices today. Our decisions are on the cutting edge of reality; they make all the difference. Skinner, more than his antagonists, would therefore agree that we should "train up a child in the way he should go" and would understand how the sins of the fathers are predictably laid upon their children "unto the third and fourth generation." Morality requires at least some regularity and predictability.

Here the mind boggles. Does not morality also require freedom to choose? How can a determined person be held morally accountable? What real choice does a determined person have?

In one sense, an absolutely determined person can have complete freedom—freedom in the practical, political sense that people care about. The opposite of determinism is not freedom in this practical sense, but indeterminism. Whether determined or not, our hypothetical person experienced a genuine, uncoerced choice—coffee or tea. Likewise, even if your vote in the next national election is predictable by an ideal reasoner who knows everything there is to know about you, you will still be free to vote for whom you wish.

The prediction does not coerce your choice, it merely describes it. Nor must we assume indeterminism in order to condemn what is evil or praise what is good. If a child were to explain, "My motives drove me to spend the bread money on candy," we would say, "Obviously, but I still hold you responsible for your choice."

To repeat, *even if* our actions were absolutely determined, we would nevertheless be free to choose consciously among alternatives, knowing that what we decide can make a great difference and that society may hold us accountable. Indeed, in Chapter 22 we will summarize scientific findings that emphasize our freedom to decide and control our own future destinies. Personal causation— the effect of what we believe and choose—is, as we will see, a tremendously important concept in contemporary psychology. What determinism denies is not the practical consequences of our inner beliefs and choices, but the philosophical idea of agent causation—that people are *ultimately* self-determining.

Divine Sovereignty and Human Responsibility

Attacks on the idea that we are self-made people—that thanks to our free will we are independently capable of righteousness— have come not only from determinists but also from several theological masterminds, among them Augustine, Martin Luther, John Calvin, and Jonathan Edwards. Together they remind us that our conception of human responsibility must not deny three attributes of God:

God's Foreknowledge

In Scripture, the selling of Joseph into slavery, the evil acts of the Pharoah, Peter's denial, Judas' betrayal of Christ, and the crucifixion are all the result of human choices that God anticipates. Such evidence moved Luther to conclude,

If we believe it to be true that God foreknows and foreordains all things; that He cannot be deceived or obstructed in His foreknowledge and predestination; and that nothing happens but at His will (which reason itself is compelled to grant); then, on reason's own testimony, there can be no "free-will" in man, or angel, or in any creature.

God's foreknowledge is, we might object, an insufficient argument for divine determinism. Surely, God is unbound by time and therefore able to see our past, present, and future. Consider, however, the implications of God's sovereignty and grace:

God's Sovereignty

Jonathan Edwards would not give so much as an inch to human free will, because to the extent that human will is indeterminant—spontaneous and free—God's plans become dependent on our decisions. But this, said Edwards, would necessitate God's "constantly changing his mind and intentions" in order to achieve his purposes. "They who thus plead for man's liberty, advance principles which destroy the freedom of God himself," the sovereign God of whom Jesus said not even a sparrow falls to the ground apart from his will. Nor is human will added to God's will such that the two together equal 100 percent. Rather, agreed Augustine, "Our wills themselves are included in that order of causes which is certain to God." God is working in and through our lives, our choices.

God's Grace

Luther argued that the bondage of the will was an essential foundation for the doctrine of grace. By ourselves, he argued, we are unable to act righteously, to have faith, and to contribute to our own salvation. All credit belongs to God. What then is left of free will? "Nothing! In truth, nothing!" insisted Luther. Calvin was just as forceful: because the term *free will* "cannot be retained without great peril, it will . . . be a great boon for the church if it is abolished."

The divine determinism assumed by the doctrines of God's foreknowledge, sovereignty, and grace is not identical with the naturalistic cause-effect determinism of Freud and Skinner. God the Creator is free to act in miraculous ways, too. Once having accepted the miracle of creation, the miracles of incarnation and resurrection become less startling, and if one accepts these—which are at the very heart of Christianity—Jesus' other miracles become plausible. To be a Christian is to accept the possibility of the miraculous. Surely a being who could create the universe could also perform smaller feats as well.

But let us also remember that God works through the created order. Thomas Aquinas argued (in the words of Michael Novak) that "grace operates (except in the rarest cases) through the ordinary contingencies and processes of nature. . . . The whole environment, the whole 'schedule of contingencies' that constitutes history, is graced." Believing in God opens one to the possibility of miracles; yet if we accept that all nature is from moment to moment sustained, ordered, and upheld by God then we no longer *need* miracles in order to "make room for God."

Whatever their differences, the concepts of absolute determinism and absolute divine sovereignty converge in affirming our dependence on forces beyond our conscious knowledge. Thus they share the problem of how to accommodate ultimate moral responsibility. If a super hypnotist were to plant an irresistible suggestion that you should commit a crime, which you then did with a sense of having chosen to do it, surely no one who knew the hidden cause of your behavior would hold you responsible. Likewise, if we understand the conditions that triggered someone's acting desirably, we tend to credit the conditions rather than the person. It is only when we are surprised by a person's heroism—when we do not expect people to behave so nobly under such circumstances—that we give special credit and honor to the hero. In a deterministic world we can judge any *behavior* as worthy or praise or blame, but it becomes more difficult to hold the *person* as ultimately responsible.

One is therefore tempted to create a gap in the schemes of natural and divine determination—to open the door to just a dash of ultimate free will, however much is needed to restore our accountability before God and before our human judicial system. God's sovereignty, we may tell ourselves, does not extend all the way down to the little things, such as what I ate for breakfast this morning. God is concerned only with the big events, the ultimate ends.

But as Edwards and the other theological masterminds recognized, this assumption of agent causation creates as many problems as it solves. A God who is detached from what you ate for breakfast (or whether you ate breakfast) is not a God who is continuously involved with all events of the creation. And consider: how are the big ends in life achieved apart from the little means? Looking back on our lives we see our path winding through countless little events and chance encounters, from our initial conception right up to the present. At any decision point we feel free, but looking back we see causation. "What I so proudly call 'myself' becomes merely the meeting place for trains of events which I never started and which I cannot stop," suggested C. S. Lewis. Or as Søren Kierkegaard noted, "Life is lived forwards, but understood backwards." Thus the apostle Paul could sense, "I . . . yet not I, but the grace of God."

So both the absolute determinist and the one who believes in God's utter sovereignty (perhaps the same person) are left baffled. To limit natural and divine powers makes little sense and only opens the door for pride in self and a judgmental attitude toward others. Yet somehow human accountability must be affirmed. Faced with this paradox of faith, we can take comfort in remembering that we cannot expect to comprehend fully the wisdom and justice of a being whose cognitive stage is infinitely beyond our own. Our situation is like that of someone stranded in a deep well with two ropes dangling down. If we grab either one alone we sink still deeper into the well. Only when we hold both ropes at once can we climb out, because at the top, beyond where we can see, they come together around a pulley. Grabbing only the rope of determinism or of human responsibiity plunges us to the bottom of a well. So instead we grab both ropes, without yet understanding how they come together. In doing so, we may also be comforted that in science as in religion, a confused acceptance of irreconcilable principles is sometimes more honest than a tidy oversimplified theory that ignores half the evidence. (Remember that advocates of agent causation have no trouble explaining our responsibility, but do face a different mystery—how God could accomplish divine purposes while granting us freedom to do as we choose.)

We also do well to remember both ropes in our everyday attitudes—by viewing ourselves as free and responsible agents and others as influenced by their biology, their past experience, and their current situation. Such a view has the effect of cultivating within us the practical fruits of self-discipline and self-initiative, while being more understanding of the forces that constrain others. Scripture, too, tends to adopt the perspective of self as free and other as caused. When the Bible addresses us directly it emphasizes our responsibility for our failings. When talking to us about others, especially the poor and disadvantaged, it frequently advocates the complementary perspective: do not judge; act with compassion toward the oppressed; take the beam out of your own eye before worrying about the motes in others; let judgment begin with the house of the Lord.

Are we determined or free? Christian psychologists who assume absolute determinism struggle to rationalize human responsibility; those who assume agent causation have solved the problem of human responsibility but struggle to accommodate natural causation and divine sovereignty in human affairs. Nevertheless, on this much both camps agree: in the fabric of contemporary psychology and Christian doctrine, natural order and human responsibility are warp and woof.

For Further Reading

Basinger, D., & Basinger, R. eds. (1986). *Predestination and Free Will: Four Views of Divine Sovereignty and Human Freedom.* Downers Grove, Ill: InterVarsity Press, 1986.
    A debate among four Christian scholars, each of whom presents his own view, which is followed by comments from the other three.

MacKay, D. M. *The Clockwork Image.* London: InterVarsity Press, 1974.
    In this intriguing 111-page book, scientist-philosopher MacKay

argues that Christian assumptions about human nature and human responsibility are valid even if our behavior is as determined as a clockwork.

R. Bufford. *The Human Reflex: Behavioral Psychology in Biblical Perspective*. New York: Harper & Row, 1981; M. Cosgrove. *B. F. Skinner's Behaviorism: An Analysis*. Grand Rapids, Mich.: Zondervan, 1980; M. S. Van Leeuwen. *The Person in Psychology: A Contemporary Christian Appraisal*. Grand Rapids, Mich.: Eerdmans, 1985.

Christian critiques of B. F. Skinner.

Chapter 13

MEMORABLE MESSAGES

As Paul was long preaching [Eutychus] sunk down with sleep, and fell
down from the third loft, and was taken up dead.

ACTS 20:9, KJV

A young couple, Martha and Leon, happily file out
from Sunday worship at Faith Church, congratulating the pastor
for the fine message on Christian love. Later that week when her
friend Sally, who was ill on Sunday, asks her about the sermon,
Martha can recall little of its content. Perhaps, Martha surmises,
she is just upset and distracted by how unloving Leon has been
lately.

Is this typical or atypical of the impact of sermons? Those of us
who teach or preach become so easily enamored of our spoken
words that we are tempted to overestimate their power. Ask college
students what aspect of their college experience has been most valuable or what they remember from their freshman year, and few
will recall the brilliant lectures given by their faculty.

Would the same be true of people reflecting on their church
experience? An award-winning study by the University of California
psychologist Thomas Crawford indicates that sermons sometimes
have surprisingly little impact. Crawford and his associates went to
the homes of people from twelve churches shortly before and after
they heard sermons opposing racial bigotry and injustice. When
asked during the second interview whether they had heard or read
anything about racial prejudice or discrimination since the previous
interview, only 10 percent spontaneously recalled the sermon.
When the remaining 90 percent were asked directly whether their

Does the Worshiper:

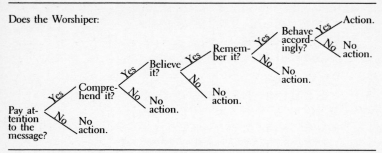

Figure 5: Requirements for an effective sermon. Adapted from the writings of Yale University social psychologist William McGuire.

minister "talked about prejudice or discrimination in the last couple of weeks," more than 30 percent denied hearing such a sermon. So it is hardly surprising that the sermons had little impact on racial attitudes!

When you stop to think about it, the preacher has so many hurdles to surmount that it's no wonder preaching so often fails to affect our actions. As Figure 5 indicates, the preacher must deliver a message that not ony gets our attention but is understandable, persuasive, memorable, and likely to compel action.

Out concern here is with neither theological content nor oratorical style, but with how to create and receive a memorable, persuasive message. What factors make for effective communication? How might ministers apply these factors in the construction of more potent messages? For that matter, how might any of us who teach, speak, or write do so with greatest effect? Finally, what can we laypeople do to receive maximum benefit from what we hear and read? Recent research has revealed five principles that help us answer these questions.

1. Vivid, Concrete Examples Are More Potent Than Abstract Information

Human judgments and attitudes are often more swayed by specific illustrations than by abstract assertions of general truth. Research studies show that a few good testimonials usually have more impact than statistically summarized data from dozens of people. One University of Michigan study found that a single vivid welfare case had more impact on opinions about welfare recipients than did factual information running contrary to the case. Another found that student impressions of potential teachers were influenced more by a few personal testimonies concerning the teacher than by a comprehensive statistical summary of many students' evaluations.

No experienced writer will be surprised by this principle. As William Strunk and E. B. White asserted in their classic, *The Elements of Style*, "If those who have studied the art of writing are in accord on any one point, it is on this: the surest way to arouse and hold the attention of the reader is by being specific, definite, and concrete. The greatest writers—Homer, Dante, Shakespeare—are effective largely because they deal in particulars." Preachers and teachers should do the same, and so should we listeners, by conjuring up our own examples when the speaker begins to get abstract.

However, a sermon is never just a string of unrelated examples; the preacher aims to communicate a basic point. We might say that theological truth is to a good sermon what the base of an iceberg is to its tip. Jesus' vivid parables, for example, embodied basic truths in memorable pictures. And what pastor has not received compliments from adults for a simple but concrete children's sermon? The children may not have grasped the analogy, but the adults understood and remembered it. This illustrates the power of principle number one: vivid, concrete examples are more potent than abstract information.

2. Messages That Relate to What People Already Know or Have Experienced Are More Easily Remembered

Public speaking experts have long supposed this to be true. Aristotle urged speakers to adapt the message to their audiences. Experimental psychologists have confirmed the point; messages that are unrelated to people's existing ideas or experiences are difficult to comprehend and are quickly forgotten. This paragraph, from an

experiment by John Bransford and Marcia Johnson, is an example
of such an unattached message:

The procedure is actually quite simple. First you arrange things into different groups. Of course, one pile may be sufficient depending on how
much there is to do. . . . After the procedure is completed, one arranges
the materials into different groups again. Then they can be put into their
proper places. Eventually they will be used once more and the whole cycle
will then have to be repeated. However, that is a part of life.

When Bransford and Johnson had people read this paragraph as
you just did, without connecting it to anything they already knew
about, little of it was remembered. When people were told that the
paragraph was about sorting laundry, something familiar to them,
they remembered much more of it—as you probably could now if
you reread it.

A message that is hooked to some cue—something we will think
about or experience again—is more likely to come to mind in the
future. When the cue pops up, it may call to mind the message
associated with it. For example, one memorable sermon likened
American religion to waiting-room Muzak—bland and soothing. A
year after this "Sound of Muzak" sermon was preached, we found
ourselves eating dinner in a room with music softly playing in the
background. Someone noticed the music—and recalled the sermon. This illustrates principle number two: messages that relate to
what people already know or have experienced are most easily remembered.

3. Spaced Repetition Aids Memory

As every student of human learning knows well, we remember
information much better if it is presented to us repeatedly, especially if the repetitions are spaced over time rather than grouped
together. Experimental psychologist Lynn Hasher has found that
repeated information is also more credible. When statements such
as "The largest museum in the world is the Louvre in Paris," were
repeatedly presented, people rated them as more likely to be true

than when the statements had been shown infrequently. Social psychologists have uncovered a parallel phenomenon: repeated presentation of a neutral stimulus—whether a human face, a Chinese character, or a piece of unfamiliar music—generally increases people's liking of it.

Speakers can capitalize on this finding that repetition, especially spaced repetition, makes messages more memorable and appealing. When preparing a talk or sermon they might ask themselves, What do I most want people to remember from this? They can then repeat that one key idea many times. (We suspect that a little informal testing of parishioners' recall would reveal that few people can recall the main points of the last three-point sermon they heard.) Given the limitations of human memory, a sermon should probably be the embodiment of only one vigorous idea. Perhaps this could even be taken a step further: that idea should be embodied in the whole worship service—the Scriptures, music, prayers, and closing charge to the congregation. As parishioners we should look for a unifying theme, or at least identify one idea in every service that is significant for us.

Sometimes the key idea can be captured in a single statement or pithy saying that becomes the trunk of a talk or sermon, unifying the illustrative branches that grow from it. What listener can forget the refrain in Martin Luther King, Jr.'s "I have a Dream" sermon? Principle number three therefore bears repeating: spaced repetition aids retention.

4. Active Listening Aids Memory and Facilitates Attitude Change

People remember information best when they have actively processed it, that is, when they have put it in their own words. Experiments reveal that when we read or hear something that prompts a thought of our own, we will often more readily remember our thought than the information that prompted it.

Not only do we better remember information we produce ourselves, but our attitudes are also more likely to be changed by that information. Social psychologists have found that passive exposure to information, through reading or listening, has less effect on people's attitudes than information they got through active participation in a group discussion. Other research confirms that when we learn something passively our attitudes toward it usually do not change much. When we are stimulated into restating information in our own terms, we are much more likely to remember and be persuaded by it.

Preachers, teachers, and even parents may fail to recognize that their spoken words are more prominent to them (as active speakers) than to their passive listeners. Parents are often amazed at their children's capacity to ignore them. If instead of constant harping, the parent gently asks the child to restate the request ("Scott, what did I ask you to do?"), the child's act of verbalizing the request will make the child more aware of it. Mr. Rogers, the television friend of preschoolers, applies this principle by asking a question and then saying nothing for a few moments, allowing children to answer for themselves. Preachers would be well advised to do likewise, pausing after giving an instruction or raising a thought-provoking question.

As listeners we can discipline ourselves to listen actively. Taking notes on a sermon, as any serious student does in class, forces us to repeat and restate its main points. So does discussing it with someone else. William James made the point eighty years ago: "No reception without *reaction*, no impression without correlative expression—this is the great maxim which the teacher ought never to forget." James anticipated principle number four: active listening aids memory and facilitates attitude change.

5. Attitudes and Beliefs Are Shaped by Action

If social psychological research has established anything, it is, as we emphasize in Chapter 27, that our actions influence our attitudes. Every time we act, we amplify the idea lying behind what we have done, especially when we feel responsibility for having committed the act. It seems that we are as likely to believe in what we have stood up for as to stand up for what we believe. Moreover, this principle is paralleled by the biblical idea that growth in faith is a consequence as well as a source of obedient action.

The implication of this "attitudes follow actions" principle is clear: a message is most likely to stimulate faith if it calls forth a specific action. The effective talk or sermon will not leave people wondering what to do with it. It will suggest specific actions, or it will stimulate listeners to form their own plan of action. "How will 'Love your neighbor' affect you?" the speaker might ask. "Who are you going to phone or visit this week?"

These five research-based principles for constructing a memorable and persuasive message can be wedded to a variety of speaking styles and theological orientations. Just remember:

1. Vivid, concrete _____ are more potent than abstract information.

2. Messages that relate to what people already _____ or have _____ are most easily remembered.

3. Spaced _____ aids retention.

4. Active _____ aids memory and facilitates attitude change.

5. Attitudes and beliefs are shaped by _____ .

And if you really want to remember these principles, look away and repeat them in your own words. Better yet, tell someone else about them, or pick out one or two and think about how you might apply them to the next sermon you prepare or hear.

For Further Reading

Cialdini, R. B. *Influence: How and Why People Agree to Things.* New York: Morrow, 1984. Also available as *Influence: Science and Practice.* Glenview, Ill.: Scott, Foresman, 1985.
 The most informative and engaging book ever written on what social psychologist Cialdini calls the "weapons of influence." Although Cialdini aims to forewarn us how persuaders may try to manipulate us, he also can be read by people wishing to become more persuasive.

Chapter 14

TO ERR IS HUMAN

What . . . then is man? How strange and monstrous! . . . Depository of truth, yet a cesspool of uncertainty and error. . . . Who will unravel this tangle?

BLAISE PASCAL
PENSEES, 1670

We psychologists are fond of noting contradictions in the accumulated cultural wisdom. We chuckle when noting that almost every conceivable proposition about human nature has at some point been argued by someone. Thus we have proverbs for many occasions, such as for times when "absence makes the heart grow fonder" and when "out of sight (becomes) out of mind." From the sages of the ages we similarly get a confusing portrayal of human nature. We know from Juvenal, the Roman satirist, that people "hate those who have been condemned" and from Ralph Waldo Emerson that "the martyr cannot be dishonored." We can base our persuasive appeals on the assumption of Shakespeare's Lysander that "the will of man is by his reason sway'd," or we can follow the advice of Lord Chesterfield: "Address yourself generally to the senses, to the heart, and to the weaknesses of mankind, but rarely to their reason." Thus we see the need for a science of human nature that will winnow fact from falsehood. Yet we know that whether we find that togetherness or separation most intensifies attraction, whether people more often despise or honor those who suffer, whether reason or emotion is generally more persuasive, there will be someone who anticipated our finding. In hindsight, almost any finding—or its opposite—can seem like obvious common sense.

True enough, and well worth remembering. But lest we falsely presume that our age has a corner on truth, we do well to recognize the insights of keen observers of human behavior in centuries past. These were people who without modern techniques of research and analysis observed the same sorts of motives and behaviors that we observe today and anticipated some of psychology's current themes. One such theme is the puzzling mixture of human wisdom and human foolishness. We today better understand how, as the seventeenth-century philosopher-mathematician Pascal put it, "Man's greatness lies in his power of thought." We marvel at the capabilities of our brains; we are awed by the seemingly limitless capacities of our visual memory; we extol our abilities to solve problems and learn language.

Yet, like Pascal, we also have been bemused and even startled by our capacity for error, illusion, and self-deceit. Because, as the psalmist recognized, "no one can see his own errors," psychologists have devoted much energy of late to revealing our most common errors. Several of these were anticipated in Francis Bacon's remarkably perceptive *Novum Organuum*, first published in 1620. Bacon, a Christian statesman and philosopher who popularized the idea that science explored God's "book of nature," identified several "idols" or fallacies of the human mind:

Finding Order in Random Events

"The human understanding, from its peculiar nature, easily supposes a greater degree of order and equality in things than it really finds." In experiment after experiment, Bacon's hunch has been confirmed; people readily see correlations or cause-effect links where there are none. Thus they all too readily make sense out of nonsense, by believing that astrological predictions predict the future, that their favorite gambling strategies can defy the laws of chance, or that superstitious rituals will bring good luck.

Overconfident Judgments

So consistent is our human tendency to overestimate the accuracy of our judgments that some researchers refer to this phenomenon as "cognitive conceit." For example, if people's answers to a factual question—such as, "Which is longer, the Panama or the Suez Canal?"—are 60 percent of the time correct, they will typically *feel* 75 percent sure. Even when they feel 100 percent sure, they still err about 15 percent of the time on such questions.

The available evidence indicates that the overconfidence phenomenon extends to scientists as they evaluate their own theories, to clinicians as they diagnose their psychologically troubled clients, and to theologians as they expound their doctrines. Bacon's words may be dated but his analysis still rings true: "Some men become attached to particular sciences and contemplations, either from supposing themselves the authors and inventors of them, or from having bestowed the greatest pains upon such subjects, and thus become most habituated to them."

The Confirmation Bias

One explanation for our persistent overconfidence is our tendency to search for and to recall instances that confirm our ideas. We are eager to verify our beliefs, less eager to seek evidence that might refute them. Bacon foresaw how a confirmation bias could maintain superstitious beliefs: "All superstition is much the same whether it be that of astrology, dreams, omens, retributive judgment, or the like, in all of which the deluded believers observe events which are fulfilled, but neglect and pass over their failure, though it be much more common."

The Biasing Power of Preconceptions

"The human mind," said Bacon

resembles those uneven mirrors which impart their own properties to different objects. . . . The human understanding, when any proposition has been once laid down (either from general admission and belief, or from the pleasure it affords), forces everything else to add fresh support and confirmation.

Dozens of recent experiments support his argument. As we noted in Chapter 3, "Should There Be a Christian Psychology?" and

Chapter 10, "Through the Eyes of Faith," it now appears that one of the most significant facts about our minds is the extent to which our preconceived notions bias how we view, interpret, and remember information. To believe is often to see and remember what we believe.

The implications of this principle extend to our understanding of Scripture. "The Bible always comes interpreted," notes church historian Martin Marty. "Oppressors and the oppressed, people in power and people out of power, Baptists of more sorts than one— all these read the book differently."

The Persuasive Power of Vivid Information

In the preceding chapter we noted that, when confronted with a compelling anecdote, people are often strangely insensitive to statistical information indicating that the anecdote is but an exception rather than the rule. Before buying his new car George consults the *Consumer Reports* survey of car owners and finds high ratings given to the Ford Escort he is considering purchasing. Hearing of his intention, Linda moans, "Oh no, not one of those! My roommate had an Escort and it was in the garage for one problem after another—cost her hundreds of dollars." The research (and our own experience) suggests that George will not do what, rationally, he ought to do—merely increment the *Consumer Reports* survey by an iota of one more car buyer. The vividness of the testimony makes it hard to forget and discount. Likewise, vivid terrorist acts in early 1986, although harming fewer than one in a million persons in Europe, caused fearful Americans to shun travel to Europe in favor of more dangerous vacations on American highways. As Bacon forewarned us,

The human understanding is most excited by that which strikes and enters the mind at once and suddenly, and by which the imagination is immediately filled and inflated. It then begins almost imperceptibly to conceive and suppose that everything is similar to the few objects which have taken possession of the mind, whilst it is very slow and unfit for the transition to the remote and heterogeneous instances by which axioms are tried as by fire.

Bacon's thoughts about thinking, and their echos in the new thinking about thinking, have several implications. For psychologists, knowing our vulnerability to error beckons us not to disparage psychological science, as some would have us do, but to restrain our unchecked speculations. Aware that we can conceive and defend almost any theory, we must be candid about our presuppositions and check our theories against the data of God's creation. To appreciate the unreliability of unchecked intuition (that sixth sense that tells us we are right, whether we are or not) is to admit that we need to do science—to wed creative intuition with systematic observation. This was Bacon's conclusion as well:

Our method and that of the sceptics agree in some respects at first setting out, but differ most widely, and are completely opposed to each other in their conclusion; for they roundly assert that nothing can be known; we, that but a small part of nature can be known, by the present method; their next step, however, is to destroy the authority of the senses and understanding, whilst we invent and supply them with assistance.

The accumulating research on human error also beckons us to a personal humility. It helps us understand why Jesus admonished us not to judge. We can easily wrong people by our overconfident judgments—that John's depression stems from his demanding parents, or that the quiet woman next door harbors suppressed hostility. Nor need we feel intimidated by other people's cockiness, least of all by dogmatic heresy hunters who are so absolutely sure they are right that they dare to practice spiritual ventriloquism—by putting their words into the mouth of God and believing the voice they throw to be the word of the Lord. When we make our own words our absolute truth, then, said the theologian Karl Barth, we have made an idol out of our religion. We have forgotten that we are not gods, but finite men and women who peer at reality in a mirror dimly. Oliver Cromwell's 1650 plea to the Church of Scotland is worth hearing over and again: "I beseech ye in the bowels of Christ, think that ye may be mistaken." The one belief of which we cannot be overconfident is the conviction that some of our beliefs contain error.

Although that may sound threatening, it should actually be re-assuring. For it means that it's okay to have doubts. Doubt reveals a mind that asks questions, a humble mind, one that does not presume its own ideas to be certainties. Indeed, the intellectually honest words *belief, faith,* and *hope* acknowledge uncertainty. We do not *believe* that $3 \times 3 = 9$, or have *faith* that what goes up will come down, or *hope* that day will follow night; we *know* these things with psychological if not logical certainty. To take the leap of faith is to bet one's life on a proposition that seems probable, that makes sense of the universe, that gives meaning to life, that provides hope in the face of adversity and death. One need not await 100 percent certainty before risking a thoughtful leap across the chasm of uncertainty. One can choose to marry in the *hope* of a happy life. One can elect a career, *believing* it will prove satis-fying. One can fly across the ocean, having *faith* in the pilot and plane. To know that we are prone to error does not negate our capacity to glimpse truth, nor does it rationalize living as a fence straddler. Sometimes, said the novelist Albert Camus, life calls us to make a 100 percent commitment to something about which we are 51 percent sure.

For Furthur Reading

Myers, D. G. *The Inflated Self: Human Illusions and the Biblical Call to Hope.* San Francisco: Harper & Row, 1980.
How do we form and maintain false beliefs? How might these illusory thinking mechanisms create erroneous beliefs in para-normal phenomena? in the quick cures of pop psychology? in superstitious religious claims? And how can we nevertheless attain a genuine faith? This book suggests some answers.

Nisbett, R., and L. Ross. *Human Inference: Strategies and Shortcomings of Social Judgment.* Englewood Cliffs, N.J.: Prentice-Hall, 1980.
Is it possible for a book that summarizes psychological research on reasoning to tickle your funny bone? This book will.

Chapter 15

SUPERSTITION AND PRAYER

"The prayer of the heart is the prayer of truth. It unmasks the many illusions about ourselves and about God and leads us into the true relationship of the sinner to the merciful God."

HENRI J. M. NOUWEN,
THE WAY OF THE HEART, 1981

The evidence is persuasive: we humans are persistently inclined to

- perceive relationships where none exist (especially where we expect to see them),
- perceive causal connections among events that are only coincidentally correlated, and
- believe that we are controlling events that we are not.

The laboratory investigations that established these phenomena have now been extended to gambling behavior, stock market predictions, clinical assessments of personality, superstitious behavior, and beliefs about ESP. People easily misperceive their behavior as correlated with subsequent events when it isn't, and thus they easily delude themselves into thinking that they can predict or control remote events.

There is every reason to suspect that illusory thinking extends to people's beliefs regarding the power of their petitionary prayers. One can hardly imagine a more perfect arena for the emergence of illusory correlations and illusions of control. Consider, too, the illusory thinking mechanisms noted in the preceding chapter—our propensity to find order in random events, to view and interpret events guided by our preconceptions, to search for and recall instances that confirm rather than disconfirm our beliefs, to be more struck by vivid anecdotes than by statistical reality. Such processes guarantee that whether our prayers change the course of events or not, we will be tempted to think they do.

If that sounds revolting and heretical, it may be reassuring to remember that warnings about false prayer come more often from believers than from skeptics. There was no stronger critic of false piety than Jesus himself. If it is heretical to think too little of our own powers, it is more heretical to think of God as a sort of celestial Santa Claus who grants our wishes if we are good, or as a cosmic vending machine whose levers we pull with our petitions. Moreover, as Henri Nouwen suggests, clearing the decks of some of the false gods of popular religion may prepare our hearts for the God of the Bible, by moving us away from preoccupation with ourselves and toward God-focused prayers—prayers of adoration, praise, confession, thanksgiving, dedication, and meditation, as well as prayers of petition.

Every belief system has those who exploit and embarrass it. Christianity is no exception. An unsolicited newspaper from "evangelist" Don Stewart promises miracles to those who will tell him their needs so he can relay them to God. One testimonial explains that "The last time I sent $10 to you to be used for God's work, God blessed me with an unexpected check in the amount of $159.74. I just can't outgive God." (One wonders, then, why the letter writer doesn't put the $159.74 back on Stewart.) Or ponder the heresy implicit in the prayer motive of American television evangelist and presidential aspirant Pat Robertson, who asked God to steer Hurricane Gloria away from his Virginia Beach television headquarters: "I felt that if I couldn't move a hurricane, I could hardly move a nation." Or consider General George S. Patton's order that all chaplains pray for an end to rains that had immobilized his troops during the winter of 1944:

General Patton: Chaplain, I want you to publish a prayer for good weather. I'm tired of these soldiers having to fight mood and floods as well as Germans. See if we can't get God to work on our side.

Chaplain O'Neill: Sir, it's going to take a pretty thick rug for that kind of praying.

General Patton: I don't care if it takes the flying carpet. I want the praying done.

The resulting prayer, which was distributed by the U. S. Army with Patton's Christmas greetings, called upon God

to restrain these immoderate rains with which we have had to contend. Grant us fair weather for Battle. Graciously harken to us as soldiers who call upon Thee that, armed with Thy power, we may advance from victory to victory, and crush the oppression and wickedness of our enemies, and establish Thy justice among men and nations. Amen.

From such uses of prayer it is a small step to anthropologist Bronislaw Malinowski's definition of superstitious magic—"acts which are only a means to a definite end expected later on." Malinowski contrasted such acts with religious acts—"acts being themselves the fulfillment of their purpose." Magic is a manipulative technique; worship is an end in itself, something intrinsically worth doing. When religion is sold as magic, using exaggerated testimonials that falsely portray faith as a route to problem-free health, wealth, and success, the convert is set up for disillusionment. Exaggeration produces doubts in those who don't get the results they expect and provokes a sense of inferiority in listeners who compare their lives with the glamorous successes of their fellow believers. If we repeatedly hear testimonies saying, "God healed all my woes," but find that our woes remain, we may either feel guilty at our seeming lack of faith or we may begin to wonder if Christianity is a farce.

But surely God does care for us, more than we can ever know. The God of the Bible is a personal being who bids us to ask, that we may receive. Although God does not promise that we will be spared sorrow, humiliation, misfortune, and death, the biblical God does offer a perspective from which to view such events and a hope that out of defeat and suffering we, like Jesus, may gain new life. How then can we separate genuine prayer from false prayer, real

faith from counterfeit faith, true Christianity from its glib carica-
tures?

One proposal has been to put prayer to the test. After all, some
popular claims for prayer are stated in a straightforward, empirical
manner. Prayer is said to produce healings, money, better weather,
parking places on busy streets, and even better grades on exams.
In a prayer-test controversy that raged in Britain during 1872–73
several scientists proposed an experiment: why not test the efficacy
of prayer as we test any other remedy? Identify a group of patients
who are suffering from a disease, administer the remedy to half,
and see if it makes a difference. If that sounds offensive to those
who remember Jesus' admonition "not to put the Lord your God
to the test" then why not examine the efficacy of prayers that have
been spontaneously uttered? The British scientist Francis Galton,
who sought to quantify everything from intelligence to female
beauty, collected mortality data on groups of people who were the
objects of much prayer—kings, clergy, missionaries—and found
that they lived no longer. Moreover, the proportion of stillbirths
suffered by praying and nonpraying expectant parents appeared to
be identical.

There are additional reasons why few Christians would put much
stock in a prayer test. For example, C. S. Lewis noted that the

impossibility of empirical proof is a spiritual necessity. A man who knew
empirically that an event has been caused by his prayer would feel like a
magician. His head would turn and his heart would be corrupted. The
Christian is not to ask whether this or that event happened because of a
prayer. He is rather to believe that all events without exception are *answers*
to prayer in the sense that whether they are grantings or refusals the prayers
of all concerned and their needs have all been taken into account.

Moreover, in analyzing the aftermath of any given granting or
refusals believers and skeptics are not going to settle their differences
through experience, for both reason, "Heads I win, tails you lose."
If the thing prayed for happens, believers see this as one more

proof that petitionary prayers work; if it doesn't, this indicates that God's will, taking everyone's known and unknown needs into account, is otherwise. In either case, the prayer is answered. Nonbelievers' reasoning is nearly the mirror image of this: if the prayed-for event happens, they see natural causes that led up to it ("It would have happened anyway"), and if it doesn't happen, that is one more proof that petitionary prayers don't work. Thus no amount of experience is likely to convince either the one who views life through the eyes of faith or the one who views life through the eyes of unbelief.

The prayer test challenge can, however, stimulate us to clarify our understanding of prayer. To believe in and wholeheartedly engage in petitionary prayer, must we agree that prayer disturbs nature's events in statistically verifiable ways? Job's experience reminds us that God does not play favorites; the rain falls both on those who plead with God and those who do not. Still, would we be wrong to presume that, other things being equal, praying parents will have 5 percent fewer stillborn or handicapped babies than nonpraying parents?

To suppose so is to fall victim to the natural/supernatural dichotomy. In the biblical view, the "God factor" is not a mere 5 percent but 100 percent. One does not need a manipulative conception of prayer to induce God's involvement in the world; God is everywhere and at all times already involved. Thus when the Pharisees pressed Jesus for some criteria by which they could validate the kingdom of God, Jesus answered, "You cannot tell by observation when the kingdom of God comes. There will be no one saying, 'Look, here it is!' or 'There it is!' for in fact the kingdom of God is among you."

What, then, is the Christian's proper prayer?

First of all, it is a declaration of praise and thanksgiving for God's infinite goodness and an acknowledgement of sin and the need for forgiveness.

> Our Father which art in heaven,
> Hallowed be thy name.
> Thy kingdom come.
> Thy will be done on earth,
>   as it is in heaven.
> Give us this day our daily bread,
> And forgive us our debts,
>   as we forgive our debtors.
> And lead us not into temptation,
>   but deliver us from evil:
> For thine is the kingdom,
>   and the power,
>   and the glory,
>   forever.
>   Amen.

Christ's prayer, the model prayer for Christians, contains no attempt to manipulate God. It does not attempt to cajole a miserly god into doing what he would not have the goodwill and good sense to do anyway. It has the quality of a confessional statement, affirming God's nature and human dependence upon God's grace. It therefore prepares one to receive that which God by his nature is already providing. The petitions that God's will be done and that forgiveness be given for debts seek what is intrinsic to God's nature. The petition for daily bread serves to reinforce the sense of God as gracious Father, of humanity as dependent and anticipating children, and of our lives as daily saturated by God's providence. "The prayer of a Christian," J. I. Packer has written, "is not an attempt to force God's hand, but a humble acknowledgement of helplessness and dependence."

Prayer is not magic, but it *is* mystical. In quiet meditation and prayer, we sense the reality of the living God. God speaks to us and we to God. As we do so we are changed. Sinking to our kness or bowing our heads reminds us of our humble dependence. Prayers for others make us more aware of their needs. As the devotional writer William Law once observed, "there is nothing that makes us love a man so much as praying for him."

Prayer may also be viewed as a response, as an effect rather than a cause, as a time not of asking

"What are we to eat? What are we to drink? What shall we wear?" All these are things for the heathen to run after, not for you, because your heavenly Father knows that you need them all. Set your mind on God's kingdom and his justice before everything else, and all the rest will come to you as well.

Paul echoes Jesus' thoughts: "The Lord is near; have no anxiety, but in everything make your requests known to God in prayer and petition with Thanksgiving. Then the peace of God, which is beyond our utmost understanding, will keep guard over your hearts and your thoughts, in Christ Jesus." Paul urges us to petition God, and we are promised an answer: not that of scientifically provable effects, but the peace of God that satisfies the deeper cravings of our being.

Jesus himself prayed that, if it be God's will, the cup might pass. It did not, but his strength was made equal to the burden. In confessing his private longings and communing with the Father, Jesus found the grace to endure. If our Creator loves us as an all-loving parent would love a child, then we, like children, can communicate with God without ceasing. We can share even the little concerns of daily existence—anything that is worth worrying about—much as a child would do with its parents or as two intimate friends do with one another. We can surrender every corner of our lives in prayer, not with a superstitious intent of manipulating magical solutions to life's problems, but in the confidence that petitionary prayer is a means of grace whereby we will grow and be sensitized to the presence of God. To ask, what's the use of petitionary prayer? is like asking what's the use of making music, skiing, or sharing a meal with a friend; such activities, like prayer, are inherently worthwhile quite apart from any further purposes they serve.

And let us not forget prayer's multiple purposes. Through prayer we thank and praise God, we humbly confess our sin and acknowledge our dependence upon God's grace, we express our concerns, and we seek inward peace and the strength to live as God's people.

For Further Reading

Myers, D. G. "Superstition and Prayer." Part 4 in *The Human Puzzle: Psychological Research and Christian Belief*. San Francisco: Harper & Row, 1978.
    More on the subject of this chapter—more skepticism about superstitious religion and more conviction about genuine prayer.
Buttrick, G. A. *Prayer*. Nashville: Abingdon-Cokesbury, 1942.
    A Christian classic about prayer by one of the great preachers of this century.

Chapter 16

WATCH YOUR LANGUAGE

All words are pegs to hang ideas on.

HENRY WARD BEECHER,
PROVERBS FROM PLYMOUTH PULPIT, 1887

One of psychology's perennial chicken-and-egg questions is which comes first, thoughts or words? Do ideas arise first and await words to name them? Or are thoughts born of words and inconceivable without them?

As usually happens with such either/or questions, the answer seems to be both: thinking shapes language, which shapes thought:

Some thoughts precede the words used to express them. Consider: to tighten a screw, which direction do you turn it?

Very likely, you first visualized the answer without words and only then expressed your thought in words such as "clockwise," or "to the right." Likewise, many artists, composers, poets, mathematicians, and scientists achieve creative insights as images. Peak religious moments, too, are sometimes experienced inarticulately; later the person struggles to express the mystical experience within the confines of language but finds it, as the apostle Paul reported, "inexpressible."

If words are sometimes the mere containers of ideas, they are nevertheless containers that shape the thoughts poured into them.

Indeed, argued linguist Benjamin Lee Whorf, "Language itself shapes a man's basic ideas." As evidence of the power of language to shape thought, Whorf pointed to the differing conceptions of reality in those who speak different languages. Because Eskimos have a variety of words that describe snow, he argued, they can more readily perceive differences in snow that often go unnoticed by English speakers. Because the Hopi Indian language has no past tense for verbs, the Hopi people cannot so readily think about the past. Likewise, people who are bilingual will readily testify that certain concepts are available to them in one language but not the other. The language-thought relationship is why so much of education is devoted to enlarging students' vocabularies. It pays to increase your word power. As Henry Ward Beecher realized, words are pegs to hang ideas on. When trained in sign language, even chimpanzees behave with an enlarged thinking power.

Because our words influence how we think, we do well to choose our words carefully. Our labels for things affect our thoughts about them. Whether a space weapons program is termed "Star Wars" or "The Peace Shield" can subtly affect people's thoughts and feelings about it. Liberation movements recognize this power of words to shape thought. When black men were called "boys" and when women were called "girls" it was easier to think of them as unequal to white men. Recognizing that racist and sexist language undergirds racist and sexist thought, one of the first goals of a liberation movement may therefore be to change the way people talk.

What is true in other realms of life is also true of religion. Our words influence our thoughts. For example, some words reflect and reinforce our tendency to think of reality as dualistic (divided into distinct categories) rather than as a unified whole. We dichotomize supernatural and natural forces, sacred and secular truths, mental and bodily realms, spiritual and material needs. Such dualisms, which as we noted in earlier chapters are more congenial with platonic than biblical assumptions, depend on certain religious words, such as *soul*. Praying about another's soul surely reflects a concern for the person's ultimate welfare, which can only be applauded. But concern for another's soul can easily degenerate into concern for an imaginary person inside the person, while one ignores the needs of the very real person who is depressed, hurting, hungry, or lonely. By expunging words such as *soul* from our vocabulary it becomes harder to think in dualistic terms.

Similarly, we may talk of *Christian life*, as when we inquire of one another, "How's your Christian life going these days?" The very words enable us to think of the Christian life as but one aspect of life, along with one's school life, sex life, vocational life, or family life. The end result is a compartmentalized view of life that assigns a corner to religion—that concerning prayer, worship, and the like—as distinct from one's studies, one's dating and family relationships, or one's aspirations. To rid oneself of such dualistic thinking a simple first step is to avoid phrases such as *Christian life* or *spiritual life*. Even in the struggle to find alternative words, one begins to view life more as a whole, no corner of which is irrelevant to being Christian.

Or consider the adjectives that people are fond of piling up before the word *Christian*. It is not enough to be simply a Christian. One must be an evangelical Christian, a mainline Christian, a Bible-believing Christian, a born-again Christian, or even a really truly born-again Christian. One scores additional points, it seems, by piling the adjectives on top of one another. Thus we have Bible-believing, Bible-teaching Christians, and even a few really truly born-again, Bible-believing, Bible-teaching, evangelical Christians.

Such words describe but also divide. The tendency of Christians (or Muslims or any other religious group) to focus on their *differences* rather than their kinship with others illustrates a powerful phenomenon: people's self-concepts center on their distinctiveness. For example, William McGuire and his Yale University colleagues report that when children are asked to tell about themselves, they spontaneously mention how they *differ* from others. Foreign-born children are more likely than others to mention their birthplace; redheads are more likely than black- and brown-haired children to volunteer their hair color; light and heavy children are the most likely to refer to their body weight; minority children are the most likely to mention their race. The principle, says McGuire, is that "one is conscious of oneself insofar as, and in the ways that, one is different." Thus "If I am a black woman in a group of white women, I tend to think of myself as a black; if I move to a group of black men, my blackness loses salience and I become more conscious of being a woman."

This insight helps us understand why Christians so often label themselves as distinct from other Christians, especially in predominantly Christian cultures. In India, Christians are more likely to see themselves simply as "Christian" (as distinct from Hindu or Muslim) and to feel a kinship with other Christians. In the United States, where a majority of the population claims to be Christian, one's distinctive identity is more likely to be a subcategory of Christian. The result is that one begins to see most fellow Christians and certainly most fellow humans as "they" and only those within one's faction as "we."

Social psychologists have also become intrigued by a subtle but reliable *ingroup bias* phenomenon. Merely assigning people an arbitrary label that they share with certain others triggers a tendency to favor one's own group and to disparage those assigned a different label. In his novel *Slapstick*, Kurt Vonnegut illustrated the phenomenon: computers gave everyone a new middle name, whereupon all "Daffodil-11's" felt kinship with one another and distance from "Raspberry-13's."

It's a point worth remembering: the labeling of who we are— our race, sex, religious denomination, and the like—also implies a definition of who we are not. The circle that defines "us" excludes "them." Devotion to and pride in one's own ethnic heritage or school or nation—or religious group—often creates a devaluation of other ethnic groups or schools or nations or religious groups. To label oneself as one of "Paul's people" or "Apollo's people," or as fundamentalist, evangelical, mainline, or liberal can be descriptive. But it can also be devisive and a source of a spiritual pride that negates Jesus' prayer "that they may all be one; even as thou, Father, art in me, and I in thee, that they may also be in us, so that the world may believe that thou has sent me."

Other words, and the images they carry, are more helpful. To say, as the apostle Paul did, that all Christians are members of one body acknowledges and accepts differences, yet encourages us to view other parts of the body as complementary to ourselves. Each part is unique and yet all work together—unity without uniformity.

The moral: let us consider our words, for powerful ideas are hung upon them.

For Further Reading

Lakoff, G., and M. Johnson. *Metaphors We Live By.* Chicago: University of Chicago Press, 1980.
    A short, delightful book that opens our eyes to metaphors, like vision, that affect how we understand and live our lives.

Chapter 17

YOU ARE GIFTED

Now there are varieties of gifts. . . .

1 CORINTHIANS 12:4, RSV

To this point, most of our reflections have noted how psychological research complements Christian faith. Some chapters have explored intriguing parallels between what researchers are concluding and what Christians believe. Other chapters have described how psychological findings can be applied to practical concerns such as preaching and prayer.

In celebrating the complementary relationships that exist between psychology and religion we must not, however, delude ourselves into thinking that there are no conflicts. As when building a tunnel between two territories, it sometimes happens that the two ends of the tunnel simply don't connect. In Chapters 19 and 25, we will note one reason why the psychology and religion tunnels may fail to connect: the two disciplines start off guided by different underlying values.

The two fields may also fail to connect because they approach a subject with two utterly different conceptions of it. A case in point is the idea of giftedness. Consider the concept of giftedness as found in psychology and in the New Testament.

The Psychological Concept of Giftedness

A family down the block has one. So do several of our colleagues. In fact, one family we know has two of them. And each of these families knows of other families who have one: a gifted child.

All across America a great hunt is on to find more gifted children. A promotional letter from the *Gifted Children Newsletter* solicits subscriptions from parents who "have the sneaking suspicion your child is *special* in some way." And how many *are* special? Dorothy Sisk, former director of the U.S. Office of Gifted and Talented estimates that "approximately 3 to 5 percent of the school-age population . . . could be considered gifted and talented." The implication is that the other 95 percent are not gifted. And that explains why, despite the lobbying of the mostly white, upper middle-class parents of these children, most school districts find "there generally are not enough gifted children" in their town to justify special programs for their gifted children.

Nevertheless, the psychology and education of the gifted few has become something approaching a social movement these days. Five national associations for the gifted have sprung up, as have several journals and magazines. Nearly every state now has a coordinator of programs for the gifted. And the American Psychological Association has a new handbook on *The Gifted and Talented*—"an extraordinary book about extraordinary people."

Virtually everyone agrees that the gifted child movement will serve a valuable purpose if it pushes schools to treat children as individuals. Not every third grader should be taking the same spelling test and working the same math problems. Better if we can find ways to individualize instruction so that no child is bored by work that is too easy or frustrated by tasks too hard. The challenge, wrote John Gardner in his book *Excellence,* is to "provide opportunites and rewards for individuals of every degree of ability so that individuals at every level will realize their full potentialities, perform at their best and harbor no resentment toward any other level."

But the problem with the gifted child movement, say its critics, is that it does *not* affirm and stimulate individuals of every ability level and, worse, may provoke resentment or self-disparagement among those implicitly labeled "*not* gifted or talented." The gifted may get to visit computer centers, do special art and science projects, visit museums, and hear guest speakers, while the non-gifted remain in their classrooms, wondering why they were excluded.

Moreover, labels such as "gifted" and "not gifted" can be self-confirming. In experiments, teachers who are told that certain children fit such labels, or students who are led to feel competent or incompetent by receiving such labels, sometimes act in ways that make the label into a reality.

In all the hoopla over giftedness what most people miss is the arbitrariness of the concept. We forget that giftedness is only a concept, artificially defined by scores among the top 3 or 4 percent on some test of aptitude or intelligence. We begin to assume that giftedness really exists out there somewhere. We come to believe it's like red hair: children either have it or they do not.

Actually, giftedness is a decision made in the minds of those who use the word. Nothing is giftedness until someone names it that. Nature has not clustered children into well-defined groups corresponding to our value-laden labels. We, not nature, decide what is a flower and what a weed. To paraphrase Ralph Waldo Emerson, a weed is but a flower that someone decides doesn't belong in the garden.

The arbitrariness of designating what is and what isn't giftedness becomes apparent when we try to agree on a practical definition. To the Yanomamo Indians of South America, giftedness is possession of the skills of a great hunter and warrior. To Suzuki violin teachers, it is musical talent. In middle-class America, one finds almost as many definitions as articles on it. But in order to pigeonhole children as gifted or not gifted we must somehow measure their giftedness. Thus we usually reduce it to a score on a one-dimensional device that measures not artistic talent or leadership skill or physical prowess or any other way in which a particular child may be special, but IQ. Period.

The Christian Concept of Giftedness

Against this currently popular idea that only a special few are gifted stands the Christian idea of giftedness. Each individual child of God is a unique part of the body of Christ, taught the apostle Paul. One person is, so to speak, a hand, another an ear, another a toe. All such parts are essential to the functioning of the whole body. Thus *each of us is gifted*. "Having gifts that differ according to the grace given to us," admonished Paul, "let us use them."

In his letters to the early churches, Paul identifies more than two dozen different gifts, challenging his readers to consider which are theirs to give. Among them are the gifts of:

Administration—to organize and direct people toward a goal

Discernment—to distinguish truth from error, good from evil

Encouragement—to support and strengthen people

Faith—a special capacity for belief and trust in God's power

Giving—an especially generous, even self-sacrificial spirit

Hospitality—to be comfortably warm and open with strangers

Leadership—to set goals and inspire a vision

Mercy—to be deeply compassionate with hurting people

Prophecy—to proclaim God's message with authority

Service—to identify needs and give effective assistance

Pastoring—to guide, nurture, and care for people

Teaching—to communicate knowledge effectively.

Paul presumed that none of us possess all of these gifts but that all of us possess at least one or two of them. None of us are completed persons by ourselves. Rather, we find our completion as we exercise our gifts in harmony with one another. Thus not everyone in a local church needs to feel compelled to teach, but some—those who have the gift of teaching—should. Others will likewise give their gifts of hospitality, administration, or mercy. The body of Christ will therefore thrive if each of us will take time to *identify our gifts*, to say no to requests that siphon our energy into areas in which we do not feel gifted, and to say yes or even to volunteer for tasks that do harness our gifts.

Consider, too, what it would mean for a college to apply the Christian idea that we all are gifted. Instead of evaluating all faculty by the same yardstick, much as schools assess giftedness with but one yardstick, a college might encourage its faculty to identify their gifts, to say no to activities in which they do not feel a special competence, and to say yes to those activities in which they do excel. Some might direct their energies more to teaching, some more to befriending and advising students, some more to research, some more to administration. If all such activities were seen as essential to the body life of the institution, then such diversity could be celebrated. If all faculty were held responsible for developing and using their gifts to the utmost, then all could esteem one another as they affirmed each others' gifts. Excellence would be expected and rewarded, but excellence might mean different priorities for different people.

And consider finally what it would mean for schools to apply the Christian idea that all are gifted, even all children. One child might be encouraged to develop his artistic talent, another her mathematical wizardry, still others their capacity for leadership or music or mechanical tasks. There would be no need to pretend that everyone is equal or to teach children as if all were the same. Indeed, would not this Christian idea of giftedness encourage John Gardner's vision of excellence, by providing opportunities and rewards such that individuals with every sort of gift "will realize their full potentialities, perform at their best and harbor no resentment toward any other level."

For Further Reading

For additonal critique of the gifted child movement, see any of the articles noted on page 209. For a more sympathetic review of the gifted child literature, see F. D. Horowitz and M. O'Brien, eds. *The Gifted and Talented: Developmental Perspectives*. Washington, D.C.: American Psychological Association, 1985.

Van Leeuwen, M. S. "I.Q.ism and the Just Society: Historical Background." *Journal of the American Scientific Affiliation* 34 (1982): 193–201.
  Tells the story of how the values of the early mental testers influenced their conceptions of intelligence, and of who had it and who didn't.

Wagner, C. P. *Your Spiritual Gifts Can Help Your Church Grow*. Glendale, Calif.: Regal Books, 1979.
  A popular book that identifies and discusses twenty-seven spiritual gifts. Suggests how to find your own gifts and how to use them.

PART 10 / Motivation

Chapter 18

TO ACCEPT OR TO CHANGE?

O God, give us grace to accept with serenity the things that cannot be changed, courage to change the things which should be changed, and the wisdom to distinguish the one from the other.

REINHOLD NIEBUHR,
"THE SERENITY PRAYER," 1943

An affirmative answer to this great prayer begins with the wisdom to distinguish that which cannot be changed from that which can. Such wisdom is now rapidly accumulating, providing us with a clearer understanding today than ever before regarding which aspects of our behavior are easily altered and which not.

Serenity to Accept What Cannot be Changed

In some respects, we seem to be less changeable, by force of will, than has been supposed. Take body weight. It has long been presumed that obesity results from gluttony, from a failure of the will, or from a personality problem such as repressed guilt or hostility. If, indeed, being overweight stems from gluttony, then losing weight should require no more than self-discipline and a trustworthy guidebook. Believing this, would-be dieters purchased diet and fitness guides in such large numbers that at least one book of this type was on the American best-seller list every week of 1985.

Of late, however, the physiologists and biopsychologists who study hunger and obesity tell us of bodily mechanisms that make permanent weight loss frustratingly difficult. That our tendencies to slimness or obesity are genetically influenced is evidenced by twin studies and by a tendency for adopted children to resemble their biological parents more than their adoptive parents. Moreover, our body fights to maintain its "set point" weight, much as a thermostat maintains room temperature at a set point. When a person diets, and body weight begins to drop, the person's metabolic idling speed decreases too, enabling the body to maintain itself on fewer calories. If the diet continues, the body's thirty billion or so miniature fuel tanks—its fat cells—may empty, but they refuse to die. Instead, they cry out, "Feed me!" by initiating biochemical processes that trigger hunger and make the dieter more responsive to external food cues and more vulnerable to eating binges.

Although it may frustrate those who weigh more than they would like, the whole system is designed to store up energy in times of abundance that will see them through times of famine. It does so with an astonishing precision that far exceeds anything you or I could achieve by conscious calorie counting. If your weight is within a pound of what it was a year ago, you have kept your average daily energy intake and output within ten calories a day of one another. Keep everything the same and add a single carrot a day and in ten years you will have gained thirty pounds. This remarkable energy control system is yet another of the wonders concealed within the most ordinary aspects of life. In ways the psalmist could not have known, we are indeed "wonderfully made."

It therefore comes as no surprise to those who understand the bodily forces that fight to maintain weight that those who have lost significant amounts of weight on diet programs nearly always, eventually, gain it back. (Most diet program entrepreneurs will deny this, but the research findings are clear.) Knowing all this can perhaps help those who are constitutionally on the chubby (or skinny) side to gain the serenity to accept what cannot be changed. Better to be a little plump than to diet and binge, to be obsessed with food, and to live with the guilt of repeated failure.

Much the same point can now be made for other human traits. Behavior geneticist Robert Plomin reports that studies of some 4,500 twin pairs, of more than 4,000 individuals in adoptive relationships, and of some 25,000 family relatives have indicated that genetic influences account for roughly half of the variation in IQ scores. Some of the remaining variation is attributable to our enriched or impoverished early experiences, which we have no power to reverse. Although intelligence is not so fixed as one's height, neither is it something that we can freely choose and freely change.

Temperament (whether one is generally excitable, intense, and reactive or easygoing, quiet, and placid) appears to be another trait that we receive rather than choose. From the first weeks of life, some human babies and some infant monkeys are more relaxed and less fearful and irritable than others. By selectively breeding, one can, within a couple dozen generations, produce one set of fierce mice and one set of placid mice. Studies of humans reveal that during adulthood, at least, one's emotional style tends to remain fairly stable. People may mellow a bit with age, but the hot-tempered young adult will usually still be recognizable as the relatively fiery golden ager. That being the case, we perhaps had best find the serenity to accept and wisely manage our temperament— and the temperaments of those around us. Better to accept what cannot be changed than always to think, "Why don't I (or you) *choose* to react differently in such situations?"

Consider finally one's sexual orientation. The wonder here is that after years of research we remain virtually ignorant of why some people become heterosexual, others homosexual. We can now discount some formerly popular theories, and new clues hint at biological factors operating before birth, but we have no reliable way of predicting—certainly not from knowledge of your relationships with your parents, your childhood sexual experiences, and so forth—whether you are primarily attracted to members of your own or the other sex. But on this much the researchers generally do agree: one's sexual orientation seems neither willfully chosen nor easily changed. Homosexual people may struggle to ignore or deny their desires and may successfully avoid acting on their desires, but the desires seldom go away. If they try to change their sexual orientation—through effort, psychotherapy, or prayer—they find that the feelings are as persistent as those of heterosexual people. This has been the experience even of several of the founders of Christian "ex-gay" programs, including Liberation in Jesus Christ, Disciples Only, Love in Action, and Exit of Melodyland, all of whom no longer pretend not to be homosexual. The story of "Ted," one highly publicized "ex-gay" illustrates the gay–becomes–ex-gay–becomes–ex-ex-gay sequence: "We all battled with our own homosexual feelings but claimed to be 'ex-gay' by faith and waited for the day it would become a reality. . . . I have yet to know of one exclusively homosexual person making such a change." One young woman, Stephanie, explains her struggle:

I had feelings for women. I would pray about these feelings all the time. I tried to keep them in the back. It was very hard. I tried to claim Bible verses and believe that I was healed, that in Christ I was a new creation. But I kept being attracted to women I would meet.

One of these women was in the Inter-Varsity chapter. We became friends and I told her my feelings for her. She wanted to help me to be healed. She and I and another woman prayed a lot. They would pray over me. We prayed for my healing. . . .

Several more unfruitful crushes followed. I finally sought help from a Christian psychotherapy center. I was in therapy for a year and a half. During this time I attended a Catholic charismatic prayer group. The priest prayed over me. He prayed that I would be healed. I wasn't. Nothing changed.

[Finally] I began to accept myself as gay. I began to feel better about myself. I came to feel that the Lord didn't condemn me. I felt I could live a responsible life as a gay person. . . .

I wish that all the people who are so concerned about changing gay people and "healing" them could see that what gay people need is to live responsible, decent lives. We need to be able to love ourselves just as anyone else—without changing from being gay. We need to be able to live with the same standards as heterosexual Christians.

Accepting the limits on our capacity to change can be liberating. To face up to the fact that we will never be thin, an A student, unperturbable, heterosexual—or any of a myriad of other things we might like to be—can be a relief. It frees us from daily living with guilt and self-blame over not having accomplished whatever it is that we keep thinking we will do. To make peace with oneself is to be able to say that grace extends to me, just as I am. Just as I am whether I am disposed to be slim or chubby, whether my grades are high or low, whether my temperament is reactive or placid, whether my longings are heterosexual or homosexual.

Courage to Change What Should Be Changed

But is there not a danger if we say no more? Might not self-acceptance of things that cannot be changed degenerate into a helpless fatalism regarding what *can* be changed? Into an attitude that says, "Whatever will be will be; I am not responsible"?

If, indeed, one's sexual orientation is neither willfully chosen nor easily changed, are there not still ethical choices that are within one's power to make? Whether heterosexual or homosexual, one can choose to engage in promiscuous sex, to elect celibacy, or to enter into a committed, loving, long-term relationship. Sexual orientation per se does not dictate the choice, nor is it an excuse for sexual exploitation of anyone.

Impatient, coronary-prone "Type A" people may not be able to change their temperament, yet they, too, retain some freedom to change their behavior. They can slow down and relax—by walking, talking, eating more slowly, smiling and laughing more, admitting their mistakes, taking more time to enjoy life, and renewing their religious faith. Meyer Friedman and his colleagues report that when one group of San Francisco heart attack survivors did so they became half as likely to experience another heart attack (compared to those who received only standard advice concerning medication, diet, and exercise).

Our aptitude may be pretty much of a given, but what we do with it surely is not. Achievement is a product of aptitude *and* motivation, inspiration *and* perspiration. Part of the perspiration is the disciplined effort to develop one's powers of reasoning. 'The biggest truth of all" about education, noted commentator Norman

Cousins, is that "its purpose is to unlock the human mind and to develop it into an organ capable of thought—conceptual thought, analytical thought, sequential thought." The effort we devote to our education—during college and after—is under our control.

Even our weight is not entirely beyond our control. Genes and physiology play a more significant role than previously supposed, but diet and exercise are significant, too. Why else is obesity so much more common among lower-class than upper-class persons, among Americans than Japanese and Europeans, and among Americans in 1987 than those in 1900? Although graduates of weight loss programs generally fail to sustain their losses, those who gravitate to such programs may be a biased sample of overweight people—namely, those who have greater-than-average difficulty maintaining weight loss. Surveys of people not in weight loss programs indicate that there are many people who, though once overweight, have managed on their own to lose the extra pounds and keep them off. If our bodies are the "temple of the Holy Spirit" and if, as a National Institutes of Health panel concluded in 1985, health risks increase with obesity, then our prayer must indeed be to find courage to change what can be changed—by minimizing exposure to tempting food cues, by exercise, and by at least a modest, sustainable adjustment in diet. Accepting ourselves means the serenity to accept bodily dispositions that cannot be changed; but loving ourselves means also being good to our bodies. For, verily, whatsoever we do unto our bodies we do unto ourselves.

O God, give us grace to accept with serenity the things that cannot be changed, courage to change the things which should be changed, and wisdom to distinguish the one from the other.

For Further Reading

Blair, R. *Ex-Gay*. Published and distributed free by Evangelicals Concerned, Suite 1403, 30 East 60th St., New York, NY 10022.
Are you more impressed than we are by reports of people who, through faith, have gained "victory" over their homosexuality? If

so, we suggest sending for Ralph Blair's critique of "ex-gay" claims.

Homosexuals Anonymous, P.O. Box 7881, Reading, PA 19605 and Outpost, 1821 University Avenue South s-296, ST. PAUL, MN 55104.

These two Christian organizations offer information, counseling, and support to those "who have made a decision to break away from the gay life" (Outpost) and who wish "to live free from homosexuality" (Homosexuals Anonymous). (Note: Just before this book went to press, Homosexuals Anonymous founder and leader, Colin Cook, resigned after admitting to repeatedly engaging in sexually intimate behaviors with male counselees.)

Chapter 19

AND GOD SAID, IT IS VERY GOOD

Male and female he created them. And God blessed them, and God said
to them, "Be fruitful and multiply. . . ." And God saw everything that
he had made, and behold, it was very good.

GENESIS 1:27–28, 31, RSV

One of the toughest issues that Christians wrestle with
is deciding how they should feel about their own sexuality. Perhaps
they have absorbed the idea that Christianity is antisexual, that Jesus
would not be pleased with their sexual feelings, that God is not
eager for them to experience pleasure, that the Bible is a book of
"thou shalt nots." Believing such, yet also feeling their pulsating
sexual urges and bombarded with cultural models and messages
advocating uninhibited sexual expression, they may feel confused,
tense, guilty, or frightened. What, then, *is* a Christian view of
sexuality, and how does this compare with the view of sexuality
found in academic and popular psychology?

The Christian View: A Fully Human Sexuality

Some Christians have indeed talked as if Christianity views sex
or pleasure or the body as bad—as something to be ashamed of.
(In the Library of Congress classification of religious books, the
subcategory of "sex" comes just after "sin.")

Actually, as we noted in Chapter 5, Christianity affirms the body.
The Bible teaches that the created material world is good, that
human nature is a mind-body unity, and that some kind of body
will be given us at the resurrection as an essential ingredient of our
everlasting happiness. In the biblical utopia of Eden, man and
woman were refreshingly accepting of their bodies—they "were

both naked, and were not ashamed" and delighted in cleaving to each other "as one flesh."

In the Old Testament, this attitude that sexuality is a gift in which we can rejoice is celebrated in the Song of Solomon as the erotic longings of lovers for each other are reciprocated openly. "I am sick with love. . . . O that his left hand were under my head, and that his right hand embraced me!" declares the bride-to-be. "You are stately as a palm tree, and your breasts are like its clusters. I say I will climb the palm tree and lay hold of its branches," replies her ardent lover.

In this biblical view, the significance of sexual intimacy is not diminished but expanded. Sexual play is a recreational activity and much more. The mutual self-exposure and cleaving of sexual activity arise from and satisfy our need for intimate communion with a loved one. To have intercourse is, in the Hebrew language, "to know" one's partner in an especially intimate way. Sexual behavior is therefore for us humans a fundamental *social* behavior, a behavior through which we may cruelly aggress against another (as does the rapist) or feel very nearly as "one flesh" in our closeness to another.

Seen in this light, the biblical laws against promiscuity and adultery are not prohibitions against pleasure. Rather, they point us toward deeper pleasures—the delicious pleasures of affection, playfulness, intimacy, and orgasm within the lifelong unity and security that we call marriage. God therefore cautions us not against pleasure, but against pleasures that are too weak, too unsatisfying for God's favorite creatures. If the Bible advises restraint of our appetites it is not because God forbids the pleasures we were created to enjoy, but because God has better things in store for us—a more fully human sexuality. God beckons us not to trade or cheapen the life-uniting, love-renewing experience of committed sexual love for a lesser and more temporary pleasure. Donald Joy (an aptly named Christian writer on this subject) likens biblical moral teachings to an owner's manual on the care and maintenance of the human machine, provided by its maker to help us fulfill its potential.

Of this much we can be sure: God is for us. If the human benefits of any activity outweigh the costs, then God is for it. And making love—genuine, self-giving, tender, joyful, caring love—is most assuredly something that God is for.

Psychological Views

Psychological views of sexuality are mixed. To some extent, they are congenial with the Christian idea that sexual intercourse is the sign and seal of a life union. Several researchers have reported that cohabitation and the number of premarital sexual partners is correlated with later marital *un*happiness and divorce. Even better established is the finding that marriage is linked with health and happiness: married people, especially married men, tend to live longer and experience greater life satisfaction than do unmarried people. Of course, such correlations do not prove that premarital chastity predisposes a happier marriage or that marriage produces health and happiness. (For example, maybe healthy, happy people are more likely to marry.) But Christians can feel reassured that the data are not inconsistent with biblical affirmations regarding sex and marriage.

Christians concerned about the sexual ideas being taught on television, in R-rated slasher movies, and in hard-core pornography can find justification for their concerns in recent research. In addition to the well-known studies on the effects of television violence on aggressive behavior, there are now many other studies indicating that:

- Television's unreal fictional world—in which acts of assault outnumber acts of affection and sexual relations occur mostly outside of marriage—affects people's perceptions of the real world.
- Pornography has evolved over the years from nudity to explicit sexual activity to sexuality involving the humiliation, exploitation, or violent degradation of women.

- Exposure to sexual violence makes many men more likely to believe that many women enjoy rape and that rape is not a serious crime, more willing to say that they might actually commit a sexual assault if they knew it would go unpunished, and more willing actually to aggress against women in laboratory settings.
- Exposure to *non*violent sexually explicit films also can make rape seem like a more trivial crime. Moreover, viewers tend to become more accepting of promiscuity, extramarital sex, and women's sexual submission. They also often become more dissatisfied with their own less sizzling sexual experiences and with their partner. Thus researcher Dolf Zillmann concludes, "If we value interpersonal sensitivity and sound human relationships in which sexuality has a vital part, if we value the nuclear family and the institution of marriage, and if we care to see women as equals rather than as servants to men's sexual imagination, pornography has demonstrable effects that are somewhat less than wonderful."

So, in some ways, especially in its gathering of evidence, psychology has reinforced Christian ideas and concerns about sexuality. (This despite potentially skewed results from the biased sample of people who may volunteer to participate in sex surveys and sex-related experiments.) Psychology has also contributed to sexual enrichment in ways that Christians and non-Christians alike can appreciate—by helping us better understand the nature of sexual motivation, by refuting myths about homosexuality, by developing new techniques for treating sexual dysfunctions.

Sexuality is, however, one of those topics about which alert readers will be sensitive to psychologists' values and assumptions. Sometimes the values of the psychologist are obvious, as when rational-emotive therapist Albert Ellis encourages "self-gratification . . . with or without long-term responsibilities" and argues that "Unequivocal and eternal fidelity to any interpersonal commitment, especially marriage, leads to harmful consequences."

More often, the assumed values are subtly expressed, as when people are urged not to repress or suppress their "natural" sexual urges, or when textbook authors, therapists, and advice columnists simply assume that sexual intimacy outside of marriage is normal and healthy. Diana Baumrind, a University of California Berkley developmental psychologist, worries that adolescents may interpret "value-free" sex education as meaning that sexual intercourse is for them a harmless recreational activity. Such a conclusion would be unfortunate, she feels, because "promiscuous recreational sex poses certain psychological, social, health, and moral problems that must be faced realistically."

In Chapter 3, we noted that values are also expressed in psychologists' labels. Labels describe behavior, but they also subtly evaluate it. When sex researchers label sexually restrained individuals as "erotophobic" or as having "high sex guilt," they both describe and evaluate such tendencies. Whether we label sexual acts we do not practice as "perversions," "deviations," or as an "alternative sexual life style" says something about our underlying attitudes.

The values of some psychologists can also be discerned from their reactions to the 1986 report of the Attorney General's commission on pornography. The commission did not label "soft-core" magazines, such as *Playboy*, as pornographic, nor did it recommend outright censorship. But, not surprisingly in view of the new research on pornography's effects, it did express strong concern about the growing availability of pornography and its effects on women's vulnerability to sexual harrassment and abuse. Although many feminist and religious leaders welcomed the report, the press preferred to quote contemptuous psychologists, such as Ted McIlvenna, president of the Institute for the Advanced Study of Sexuality: "Sexologists should universally view this report with disdain. . . . It is totally an erotophobic report. . . ." The problem, some of these psychologists complained, was that the commission took the psychological research on pornography *too* seriously—an interesting reversal from the more common complaint that government bodies have ignored research.

So, in welcoming sexual advice and sex research Christians should be critically alert to psychologists' values. On the other hand, we should also counter the unbiblical idea that sexual relations are bad or that passion is something to be ashamed of. In the final analysis, we Christians are not so opposed to promiscuity and pornography as we are *for* the intense closeness and the pleasureful sense of union that occurs when wife and husband express and renew their love. In such a relationship sexuality has the potential to become all that the Bible envisions its being: procreational, recreational, and deeply affectional.

Let's therefore remember: our sexuality is a gift that the Lord has made. Let us rejoice and be glad in it, knowing that God said, it is *very* good.

For Further Reading

Scanzoni, L. D. *Sexuality.* Philadelphia, Pa.: Westminster Press, 1984.
       A positive, helpful, and lucid little book about female sexuality by a Christian writer who has also coauthored a leading text on marriage and family.
Smedes, L. B. *Sex for Christians: The Limits and Liberties of Sexual Living.* Grand Rapids, Mich.: Eerdmans, 1976.
       Seminary professor Smedes offers a thoroughly biblical and unprudish look at the meaning and practice of sexuality. Highly recommended reading.

Chapter 20

THIS WAY TO HAPPINESS

How to gain, how to keep, how to recover happiness, is in fact for most men at all times the secret motive of all they do, and of all they are willing to endure.

WILLIAM JAMES,
VARIETIES OF RELIGIOUS EXPERIENCE, 1902

Happiness—that delicious feeling of well-being and joy. What does it mean for our lives? How can we attain it?

Have you noticed how your states of happiness and unhappiness color everything else? Researchers have found that when we are in a happy mood we see the world as friendly and nonthreatening. We make decisions easily. We recall the good times and forget the bad. Let our mood turn gloomy and soon enough we'll find reasons for it: our relationships, our self-image, and our prospects for the future suddenly seem depressing.

What is more, happy people are helpful people. In experiments, those who have had a mood-boosting experience become more generous and compassionate. If made to feel successful and intelligent they are more likely to volunteer as a tutor. If they have just found some money in a phone booth they are more likely to help someone pick up dropped papers. If they have just listened to a Steve Martin comedy album they are more willing to loan someone money.

So, being in a good mood triggers happy thoughts and memories and predisposes us to spread happiness to others. How, then, can we find happiness? Psychologists who have studied happiness offer several hints.

1. Break the Vicious Cycle of Negative Thinking

When faced with severe adversity or loss, being depressed is a normal and appropriate response. But sometimes people react even to little problems by doubting and disparaging themselves. Their negative mood now triggers more negative thoughts: "I'm no good," "People don't like me," "No one appreciates the work I do." And the withdrawal and complaining that accompany such thoughts irritates others, which further worsens the unhappy person's predicament.

To break this vicious cycle of misery, psychologists often advise people to work at reversing their negative thinking. Keep a diary of daily successes, noting what you did to make them possible. Make negative self-talk more positive: not "I'll never get this done," but "One step at a time—I can handle it."

Forcing ourselves also to *act* in more positive ways—offering a compliment, asserting oneself—can help, too. When we act *as if* we are happy and confident, we may become more so. Silly as it may seem, even a smiling expression can sometimes break the cycle of misery. Try it. Fake a big grin. Can you feel the difference?

The participants in dozens of recent experiments could feel the difference. When induced to make a frowning expression while electrodes were attached to their faces—"pull your brows together, please," the researcher might instruct—the people reported *feeling* a little angry, and their heart rates and skin temperatures actually went up slightly (as if they really were "hot under the collar"). Those induced to smile felt happier and found cartoons more humorous. When we put on a happy face, our outlook seems to brighten.

2. Set Realistic Expectations

Ever-rising expectations mean never-ending dissatisfaction. That simple principle helps explain a surprising but consistent finding: although most Americans believe that 10 or 20 percent more income would make them happier, those who have or who gain the additional income are in the long run no happier. Beyond enough money to afford life's necessities, more money does not buy more happiness. Since the 1950s, Americans' buying power (their income adjusted for taxes and inflation) has doubled; their self-rated happiness is unchanged. When a 1982 Roper survey invited Americans to rate their satisfaction with thirteen different aspects of their lives (their marriage, their spouse, their schooling, and so forth) dissatisfaction was greatest over "the amount of money you have to live on." Thus more money has not even provided contentment with money. Even those who have won state lotteries report their overall happiness unchanged.

Why? Because once we have adapted to any level of attainment we require something better yet to give us another surge of pleasure. And having adapted to good times, we then feel the pain of deprivation if forced to adjust downward, even if only to levels that once were gratifying. Black and white television, once a thrill, seems dull after we have adapted to color.

The principle can be generalized: every desirable experience—being engrossed in one's work or study, passionate love, feelings of spiritual vitality, pleasure in achievement or financial gain—is transient. Freud saw this: "When any situation that is desired by the pleasure principle is prolonged, it only produces a feeling of mild contentment." So, too, have researchers who study the phenomena that psychologists know as "adaptation-level" and "opponent processes," which imply that human emotions inevitably fall back from elation (or despair) toward neutrality. Social psychologists Phillip Brickman and Donald Campbell explain: "While happiness, as a start of subjective pleasure, may be the highest good, it seems to be distressingly transient." The Roman statesman Seneca put it simply: "No happiness lasts for long."* This being the earthly reality, what greater image of heaven can there be then that suggested by

---

* An appreciation of the transience of peak emotions is vital for newlyweds: even in the best of marriages, passionate love inevitably settles within a year or so to a more tranquil afterglow. Those who do not understand the odds against eternal passion may therefore decide that something is wrong and that the relationship should be terminated; those who understand may instead relax and enjoy the quieter feelings of satisfaction, security, and contentment.

C. S. Lewis at the end of his *Chronicles of Narnia*—a place where happiness is eternally expanding. For all the adventures in Narnia and on Earth were but the cover and title page before "Chapter One of the Great Story, which no one on earth has read: which goes on forever: in which ever chapter is better than the one before."

There is another reason why new attainments do not bring lasting happiness: when our achievements rise, so do our standards of comparison. We still look at people who have what we want, and we continue to feel envious or dissatisfied. Advertisers exploit this by bombarding us with images of happy people whose cars and clothes are far superior to our own. But consider how we can use the effect of comparison to set expectations that increase our contentment.

3. Compare with Those Less Fortunate

The psychologist Abraham Maslow once noted that to appreciate your blessings "all you have to do is go to a hospital and hear all the simple blessings that people never before realized *were* blessings—being able to urinate, to sleep on your side, to swallow, to scratch and itch." In one experiment, students viewed depictions of poverty or were asked to imagine tragedies such as being burned and disfigured. Afterward, these students expressed greater satisfaction with their own lives. Comparing with those who have more may breed envy, but comparing with those who have less breeds contentment. So count your blessings, name them one by one.

4. Take Pleasure in the Moment

There is an old Chinese story about a dog who discovers that happiness lies in his tail: "When I chase my tail I never catch it. But when I go about my business, it follows me." The moral is that to achieve happiness we must abandon our striving for a happiness that lies yet ahead of us.

To be happy is to take pleasure in the moment. "The streams of small pleasures fill the lake of happiness." observed Benjamin Franklin. Sipping a cup of tea while reading a good book, taking joy in a problem solved, delighting in a child's excitement, being gratified by the affirming smile and touch of a loved one—these are moments of happiness for those not too busy to catch them as they fly by. "Martha, Martha, you are anxious and troubled about many things," said Jesus; learn from Mary, who has chosen to relax and savor these moments.

5. Living Beyond Self-Concern

Happy moments are also there for those who become absorbed in things beyond themselves. The psychologist Maslow contended that self-actualized people tend to be loving, caring people, most of whom focus their energies on some task, which they regard as a mission in life. For us as for them, happiness is a by-product of the satisfactions and self-worth that come from involvement in things larger than ourselves.

Many such people discover their sense of meaning and missions through mystical or spiritual experience. Survey researchers report that those who pray or have intense mystical experiences report being happier than those who don't. Among the elderly, deeply religious persons tend to report greater satisfaction with life. And George Gallup reports that people who are "highly spiritually committed" are also "far happier" and more satisfied with life—68 percent rating themselves as "very happy," compared to 46 percent of those moderately spiritually committed, 39 percent of the moderately uncommitted, and 30 percent of the highly uncommitted.

We must be careful not to overinterpret these correlations between self-reported spirituality and happiness. But the spirituality-happiness link is perhaps what Jesus had in mind in one of his teachings that all four gospel writers thought important enough to record: "He who loses his life for my sake will find it."

Continual happiness is beyond the reach of life on this earth. Nevertheless, thinking and acting positively, setting realistic goals, counting one's blessings by comparing with those less fortunate, savoring pleasureable moments, living with spiritual commitment rather than self-focus—these are among the ingredients of a general sense of well-being.

For Further Reading

Argyle, M. *The Psychology of Happiness*. London: Methuen, 1987.
A comprehensive review of what psychologists have learned about the roots and fruits of happiness.

Ludwig, T., M. Westphal, R. Klay, and D. Myers. *Inflation, Poortalk, and the Gospel*. Valley Forge, Pa.: Judson Press, 1981.
Two psychologists, an economist, and a philosopher together discuss how greed has infected the American dream, how money does (and does not) relate to happiness, and how Christians can experience well-being amidst varying circumstances.

Chapter 21

A NEW LOOK AT PRIDE

We are all so blinded and upset by self-love that everyone imagines he has a just right to exalt himself, and to undervalue all others in comparison to self.

If God has bestowed on us any excellent gift, we imagine it to be our own achievement, and we swell and even burst with pride.

JOHN CALVIN,
GOLDEN BOOKLET OF THE TRUE CHRISTIAN LIFE, 1549

It is widely believed that most of us suffer the "I'm not OK—you're OK" problem of low self-esteem, the problem that comedian Groucho Marx had in mind when he declared, "I wouldn't want to belong to any club that would accept me as a member." The humanistic psychologist Carl Rogers asserted this low self-image problem when objecting to theologian Reinhold Niebuhr's idea that original sin is self-love, pretension, and pride. Wrong, objected Rogers: people's problems arise because "they despise themselves, regard themselves as worthless and unlovable."

Thirty years after the Niebuhr-Rogers exchange the self-image issue remains alive. Ironically, many Christian preachers and writers are echoing the teachings of humanistic psychology by telling us that the fundamental human problem is low self-esteem. Meanwhile, research psychologists have been amassing new findings concerning the pervasiveness of pride. Indeed, it is the older theologians such as Niebuhr, not the humanistic psychologists and their Christian popularizers, who seem best to have anticipated a phenomenon uncovered by recent research. As writer William Saroyan put it, "Every man is a good man in a bad world—as he himself knows."

Researchers debate the sources of this self-serving bias phenomenon but agree that various streams of data merge to confirm its pervasiveness. Consider:

Stream 1: Accepting More Responsibility for Success Than Failure, for Good Deeds Than Bad

Time and again, experimenters have found that people readily accept credit when told they have succeeded (attributing the success to their ability and effort), yet they attribute failure to external factors such as bad luck or the problem's inherent "impossibility." These self-serving attributions have been observed not only in laboratory situations, but also with athletes (after victory or defeat), students (after high or low exam grades), drivers (after accidents), and married people (among whom conflict often derives from perceiving oneself as contributing more and benefiting less than is fair). Self-concept researcher Anthony Greenwald summarizes, "People experience life through a self-centered filter."

Stream 2: Favorably Biased Self-ratings: Can We All Be Better Than Average?

In virtually any area that is both subjective and socially desirable, most people see themselves as better than average. Most business people see themselves as more ethical than the average business person. Most community residents see themselves as less prejudiced than their neighbors. Most people see themselves as more intelligent and as healthier than most other people. When the College Board asked high school seniors to compare themselves with others their own ages, 60 percent reported themselves better than average in athletic ability, only 6 percent below average. In leadership ability, 70 percent rated themselves above average, 2 percent below average. In ability to get along with others, *zero* percent of the 829,000 students who responded rated themselves below average, while 60 percent saw themselves in the top 10 percent and 25 percent put themselves in the top 1 percent. If Elizabeth Barrett Browning were still writing she would perhaps rhapsodize, "How do I love me? Let me count the ways."

Stream 3: The Barnum Effect

"There's a sucker born every minute," said showman P. T. Barnum. A number of experiments have given us a psychological version of that maxim. The procedure is simple: people are shown statements such as those in horoscope books ("You have a strong need for other people to like you and for them to admire you. . . . While you have some personality weaknesses, you are generally able to compensate for them. . . . At times you are extroverted, affable, sociable, while at other times you are introverted, wary, and reserved"). If told that the description is designed specifically for them on the basis of their psychological tests or astrological data, people usually say the description is remarkably accurate, *especially when it is favorable*. Negative assessments are judged less valid than flattering ones. "The Arch-flatterer," noted Plutarch, "is a man's self."

Stream 4: The Totalitarian Ego

At the University of Waterloo, Michael Ross has repeatedly found that people will distort their past in ego-supportive ways. In one experiment he exposed some people to a message about the desirability of frequent toothbrushing. Shortly afterwards, in a supposedly different experiment, these students recalled brushing their teeth more often during the preceding two weeks than did an equivalent sample of people who had not heard the message. Noting the similarity of such findings to happenings in George Orwell's *Nineteen Eighty Four*—where it was "necessary to remember that events happened in the desired manner"—Anthony Greenwald surmised that human nature is governed by a totalitarian ego that continually revises the past in order to preserve a positive self-evaluation.

Because of our mind's powers of reconstruction, we can be sure, argues Mike Yaconelli, that "Every moving illustration, every gripping story, every testimony, didn't happen (at least, it didn't happen like the storyteller said it happened)." Every anecdotal recollection told by a Christian superstar is a reconstruction. It's a point worth remembering in times when we are feeling disenchanted by the comparative ordinariness of our everyday lives.

Stream 5: Self-justification: If I Did It, It Must Be Good

If an undesirable action cannot be forgotten, misremembered, or undone, then often it is justified. Among psychology's best-established principles is that our past actions influence our current attitudes. Every time we act, we amplify the idea lying behind what we have done, especially when we feel some responsibility for having committed the act. In experiments, people who oppress someone—by delivering electric shocks, for example—tend later to disparage their victim.

Stream 6: Cognitive Conceit: Belief in One's Infallibility

Researchers who study human thinking have often observed that people overestimate the accuracy of their beliefs and judgments. As Baruch Fischhoff and others have demonstrated, we often do not expect something to happen until it does, at which point we overestimate our ability to have predicted it—the "I knew it all along" phenomenon. People also fail to recognize their vulnerability to error. (Recall from Chapter 14 the overconfidence phenomenon.)

Stream 7: Unrealistic Optimism: The Pollyanna Syndrome

Margaret Matlin and David Stang have amassed evidence pointing to a powerful Pollyanna principle—that people more readily perceive, remember, and communicate pleasant than unpleasant information. Positive thinking predominates over negative thinking. At Rutgers University, Neil Weinstein also has discerned a consistent tendency toward unrealistic optimism about future life events. Most students perceive themselves as far more likely than their classmates to experience positive events such as getting a good job, drawing a good salary, and owning a home, and as far less likely to experience negative events such as getting divorced, having cancer, and being fired.

Stream 8: Overestimating How Desirably One Would Act

Under certain conditions, most people have been observed to act in rather inconsiderate, compliant, or even cruel ways. When other people are told about these conditions and asked to predict how *they* would act, nearly all will insist that their own behavior would be virtuous. Similarly, when researcher Steven Sherman called Bloomington, Indiana, residents and asked them to volunteer three hours to an American Cancer Society drive, only 4 percent agreed to do so. Meanwhile, a comparable group of other residents were being called and asked to predict how they would react were they ever to receive such a request. Almost half claimed they would help.

There are additional streams of evidence, but the point is made: the most common error in people's self-images is not unrealistically low self-esteem, but rather self-serving pride; not an inferiority complex, but a superiority complex.

No doubt many readers are rightly objecting that this portrayal of self-image is one-sided, and that inflated egos are sometimes a crust that hides inner wounds. True enough, as we ourselves will argue in subsequent chapters. Other readers may object that the portrayal clashes with their own experience: "The people I talk to seem to put themselves down more than they flatter themselves, and I'm sometimes plagued by inferiority feelings myself." Let's see why this might be so. First, all of us some of the time *do* feel inferior, especially when comparing ourselves to those who are a step or two higher on the ladder of grades, income, looks, or popularity.

Second, some people (notably those who are depressed) tend not to exhibit self-serving bias. Most people shuck responsibility for their failures or perceive themselves as being more in control than they are or perceive themselves more favorably than other people see them; but depressed people tend to be more accurate in their self-appraisals. "Sadder but wiser," reports one prominent researcher.

Third, self-disparagement can be subtly self-serving. Putting oneself down is an effective technique for eliciting strokes from others. A remark such as "I wish I weren't so ugly" will generally elicit at least a "Come now, I know people who are uglier than you." People have also been observed to aggrandize their opponents and to disparage or handicap themselves for self-protective reasons. The coach who publicly extols the upcoming opponent's awesome strength makes a loss understandable and a win more praiseworthy.

Although self-serving bias is in some ways adaptive, it can wreak social havoc. In a long series of investigations at the University of Florida, Barry Schlenker repeatedly found that people in groups claim greater-than-average credit when their group does well on a task, and less-than-average blame when it does not. When most people in a group believe they are underpaid and underappreciated, given their better-than-average contributions, disharmony and envy surely lurk. Several surveys indicate that 90 percent or more of college faculty think themselves superior to their average colleague; when merit salary raises are announced and half receive an average raise or less, many are bound to perceive an injustice.

The abundant evidence that human reason is adaptable to self-interest parallels the Christian claim that becoming aware of our sin is like trying to see our own eyeballs. There are self-serving, self-justifying biases in how we perceive and interpret our actions, echo the researchers. Thus the Pharisee could thank God "that I am not like other men" (and we can thank God that we are not like the Pharisee). The apostle Paul must have had this self-righteous tendency in mind when he admonished the Philippians to "in humility count others better than yourselves."

Note that Paul assumed that our natural tendency is to count ourselves better than others, just as he assumed self-love when he argued that husbands should love their wives as their own bodies, and just as Jesus assumed self-love when commanding us to love our neighbors as we love ourselves. The Bible neither teaches nor opposes self-love; it takes it for granted.

The Bible does, however, warn us against self-righteous pride— pride that alienates us from God and leads us to disdain one another. Such pride is at the heart of racism, sexism, nationalism, and all the deadly chauvinisms that lead one group of people to see themselves as more moral, deserving, or able than another. The flip side of being proud of our individual and group achievements, and taking credit for them, is blaming the poor for their poverty and the oppressed for their oppression. Samuel Johnson recognized this in one of his eighteenth-century *Sermons:* "He that overvalues himself will undervalue others, and he that undervalues others will oppress them." The Nazi atrocities were rooted not in self-conscious feelings of German inferiority but in Aryan pride. The arms race is fed by a national pride that enables each nation to perceive its own motives as righteously defensive, the other's as hostile. Even that apostle of positive thinking Dale Carnegie foresaw the danger in 1936: "Each nation feels superior to other nations. That breeds patriotism—and wars."

And so for centuries pride has been considered the fundamental sin, the original sin, the deadliest of the seven deadly sins. Vain self-love corrodes human community and erodes our sense of dependence on one another and on God. If we seem confident about the pervasiveness and potency of pride, it is not because we have invented a new idea, but rather because the new findings reaffirm a very old idea.

For Further Reading

Fairlie, H. *The Seven Deadly Sins Today.* Washington, D.C.: New Republic Books, 1978.
    Essayist and social critic Fairlie brings us face-to-face with today's (and yesterday's) seven deadly sins: pride, envy, anger, sloth, avarice, gluttony, lust.
Hoekema, A. *The Christian Looks at Himself.* Grand Rapids, Mich.: Eerd-

mans, 1975.

Theologian Hoekema explores Christian resources for a healthy self-image.

Chapter 22

THE POWER OF POSITIVE THINKING

If you think in negative terms you will get negative results. If you think in positive terms you will get positive results. That is the simple fact . . . of an astonishing law of prosperity and success.

NORMAN VINCENT PEALE,
THE POWER OF POSITIVE THINKING, 1952

The preceding chapter emphasized an underappreciated truth: the potent and corrupting power of self-serving pride. But as Pascal taught, no single truth is ever sufficient, because the world is not simple. Any truth separated from its complementary truth is a half-truth. It is true that pride leads to self-sufficient individualism, the taking of credit and displacement of blame, and an intolerance of those "inferior." However, let us not forget the complementary truth about the benefits of positive self-esteem and positive thinking.

Jesus called us to self-denial: "If any man would come after me, let him deny himself and take up his cross and follow me," but not to self-rejection. Far from devaluing our individual lives, he proclaimed their value. Being created in the image of God, we are more valuable than "the birds of the air" and the other animals for whom God cares. As one young victim of prejudice insisted, "I'm me and I'm good, 'cause God don't make junk." Indeed, our worth is more than we appreciate—worth enough to motivate Jesus' kindness and respect toward those dishonored in his time—women and children, Samaritans and Gentiles, leprosy victims and prostitutes, the poor and the tax collectors. Recognizing that our worth is what we are worth to God—an agonizing but redemptive execution on a cross—therefore draws us to a self-affirmation that is rooted in divine love.

Without doubt, such feelings of self-worth pay dividends. People who feel good about themselves—who express a positive self-esteem—are generally less depressed, freer of certain ailments and drug abuse, more independent of peer pressure, and more persistent when facing tough tasks. Many clinicians report that underneath much of the human despair and disorder with which they deal is an impoverished self-acceptance, a sense that "I *am* junk."

The sharp-eyed psychology student will recognize that cause and effect are ambiguous in this correlation between misery and self-rejection. Perhaps miserable experiences cause feelings of worthlessness rather than the other way around. But experiments indicate that a lowered self-image can indeed have negative consequences. Imagine yourself being temporarily deflated by the news that you scored poorly on an intelligence test or that some people you met earlier thought you were unappealing and unattractive. Might you react as experimental subjects often have—by disparaging others or even exhibiting racial prejudice as a way to restore your feelings of self-worth? The defensive, self-righteous pride that feeds contemptuous attitudes can itself be fed by the inner turmoil of self-doubt. People who are made to feel insecure and who therefore have a need to impress others are more likely to make scathing assessments of others' work than are those who feel secure and comfortable with themselves. Mockery says as much about the mocker as the one mocked.

Positive thinking about one's potential also pays dividends. The positive thinking preachers Norman Vincent Peale and Robert Schuller would be pleased but not surprised at the breadth of psychological research that confirms the power of faith in one's possibilities. For example:

Those who believe they can control their own destiny, who have what researchers in more than a thousand studies have called *internal locus of control*, achieve more, make more money, and are better able to cope with problems. Believe that things are beyond your control and they probably will be. Believe that you can do it, and maybe, just maybe, you will.

Studies with both animals and people reveal a phenomenon of tragic resignation, called *learned helplessness*. When dogs are strapped in a harness and repeatedly given unavoidable shocks, they learn a sense of helplessness. When later they are given shocks in a situation where all they need do is leap a hurdle to escape, they fail to do so. Other animals, which have previously been able to escape shocks, face the new situation with a more positive attitude that enables them to triumph over the trauma. Likewise, oppressed people—even those who passively receive well-intentioned nursing home care—decline more rapidly than do those who are encouraged to exert some personal control over their environment. Jesse Jackson has carried this hopeful, take-control-of-your-future attitude to black youth, an attitude conveyed by his speech to the 1983 civil rights march on Washington: "If my mind can conceive it and my heart can believe it, I know I can achieve it."

Additional studies indicate that when people undertake challenging tasks and succeed, their feelings of self-efficacy are strengthened. For example, people who are helped to conquer an animal phobia may subsequently become less timid and more self-directed and venturesome in other areas of their life. Albert Bandura, a recent president of the American Psychological Association, theorizes that the key to self-efficacy is not merely positive self-talk ("I think I can, I think I can") but actual mastery experiences—tackling realistic goals and achieving them.

Advocates of *cognitive behavior therapy* are even more optimistic about the power of positive thinking. Such therapy aims to reverse the negative, self-defeating thinking that underlies difficulties such as depression. It trains people to see how their negative interpretations make them depressed and to reverse their negative self-talk.

Additional studies on *intrinsic motivation*, on *achievement motivation*, on the importance of perceived choice in studies of action-produced attitude change, and on the phenomenon of *reactance* (a motive to restore one's freedom when feeling coerced) further testify to the benefits of believing in our own possibilities. These studies also put the mainstream of recent psychological research squarely behind conceptions of human freedom, dignity, and self-control. The moral of all these research literatures is that people benefit from experiences of freedom and from being able to view themselves as free creatures rather than as pawns of external forces.

But this truth also has a complementary truth: the perils of positive thinking. One such peril is the guilt, shame, and dejection that may accompany shattered expectations. If a 1982 *Fortune* magazine ad was right in proclaiming that you can "make it on your own," on "your own drive, your own guts, your own energy, your own ambition," then whose fault is it if you *don't* make it on your own? If Barbara Smallwood and Steve Kilborn are right to say that "what you believe yourself to be, you are. . . . Believing is magic. You can always better your best," then whose fault is it if you don't progress upward from highs to higher highs? Whose fault is it if Amway President Richard DeVos was correct in explaining why so many Amway distributors fail? "Those who really want to succeed, succeed; the others didn't try hard enough." What do we conclude when, as is the case with most of us, our marriages are less than ideal, our children are flawed, our vocations less successful than we dreamed? In *Death of a Salesman*, Arthur Miller suggested that by trying too hard to win, one ultimately loses when the dream collapses. Limitless expectations breed endless frustrations. "Blessed is he who expects nothing, for he shall never be disappointed," counseled poet Alexander Pope in a 1727 letter. Life's greatest disappointments, as well as its highest achievements, are born of the most positive expectations.

A second peril of positive thinking is that one begins to live in the future rather than the present. C. S. Lewis's devilish Screwtape advised Wormwood to "fix men's affections on the Future," where nearly all vices are rooted: "Gratitude looks to the past and love to the present; fear, avarice, lust, and ambition look ahead." By so doing, Screwtape hoped to counter his enemy's ideal of the person "who, having worked all day for the good of posterity (if that is his vocation), washes his mind of the whole subject, commits the issue to Heaven, and returns at once to the patience or gratitude demanded by the moment that is passing over him." Pascal, too, saw the perils of endless ambition: "The present is never our end. The past and the present are our means—the future alone is our end. So we never live, but we hope to live—and as we are always preparing to be happy, it is inevitable we should never be so."

The third peril of positive thinking is an excessive optimism that leads to complacency about evil. In the face of a worldwide arms race, exploding population, and assaults on the environment, positive thinkers are inclined not to worry. "The pessimists have often been wrong in the past," they say, "so let's not trouble ourselves with their negative thinking." It was an optimistic we-can-do-it attitude that emboldened Lyndon Johnson to invest our weapons and soldiers in the effort to salvage democracy in South Vietnam. It was positive thinking that gave Jimmy Carter the courage to attempt the rescue of American hostages in Iran. It was possibility thinking that enabled a resolute Ronald Reagan to send troops to Lebanon in hopes of restoring peace, to support the Contra's guerilla warfare in Nicaraugua, to assume that selling weapons of death to Iran would promote moderation and reduce the number of American hostages in Lebanon.

By contrast, experiments indicate that one type of negative thinking—anxiety over contemplated failure—can motivate high achievement. (Think of the students who, fearing they are going to bomb the upcoming exam, proceed to study furiously and, not surprisingly, get A's.) To be sure, hopeless despair breeds as much apathy as does naive optimism. What we therefore need is neither negative nor positive thinking, but realistic thinking—thinking characterized by enough pessimism to trigger concern, enough optimism to provide hope.

How then can we realize self-denial without self-rejection? Self-affirmation without vain self-love? And what is a genuine Christian humility?

First, we must remember that humility is *not* self-contempt. To paraphrase C. S. Lewis, humility does not consist in handsome people trying to believe they are ugly and clever people trying to believe they are fools. Ivan Lendl and Martina Navratilova can acknowledge their greatness at tennis without violating the spirit of humility. False modesty regarding one's gifts can actually lead to pride—pride in one's better-than-average humility. Screwtape recognized this possibility in advising Wormwood to catch his prey "at the moment when he is really poor in spirit and smuggle into his mind the gratifying reflection, 'By jove! I'm being humble,' and almost immediately pride—pride at his own humility—will appear."

True humility also is not found by struggling to straddle the fence between egotistical vanity and self-hatred. Humility is more like self-forgetfulness. It is flowing with life with minimal self-consciousness, as when we become totally absorbed in a challenging task, an exciting game, or even a life mission. Dancers, athletes, chess players, surgeons, and writers often experience this kind of absorption. With it comes a satisfaction that accompanies the relinquishment of the self-conscious pursuit of happiness. Dennis Voskuil states the phenomenon in Christian terms: the refreshing gospel promise is "not that we have been freed by Christ to love ourselves, but that we are free from self-obsession. Not that the cross frees us *for* the ego trip but that the cross frees us *from* the ego trip." This leaves us free to esteem our special talents and, with equal honesty, to esteem our neighbors. Both the neighbor's talents and our own are recognized as gifts that, like our height, demand neither vanity nor self-deprecation.

Obviously, true humility is a state not easily attained. C. S. Lewis offered, "If anyone would like to acquire humility, I can, I think, tell him the first step. The first step is to realize that one is proud. And a biggish step, too." The way to take this first step, continued Lewis, is to glimpse the greatness of God and see oneself in light of it. "He and you are two things of such a kind that if you really get into any kind of touch with Him you will, in fact, be humble, feeling the infinite relief of having for once got rid of [the pretensions which have] made you restless and unhappy all your life." To be self-affirming yet self-forgetful, positive yet realistic, grace-filled and unpretentious—that is the Christian vision of abundant life.

For Further Reading

Schuller, R. *Self-Esteem: The New Reformation*. Waco Texas: Word Books, 1982.
  The famous possibility-thinking pastor offers his corrective for paralyzing negative emotions.
Voskuil, D. *Mountains into Goldmines: Robert Schuller and the Gospel of Success*. Grand Rapids, Mich.: Eerdmans, 1983.
  Voskuil offers a readable history of the positive thinking movement in American religion and goes on to describe and analyze the ministry of television preacher Robert Schuller.

Chapter 23

IS CHRISTIANITY BENEFICIAL OR HAZARDOUS TO YOUR MENTAL HEALTH?

Do not be anxious about your life.

MATTHEW 6:25, RSV

Consider Francis, the popular son of a wealthy textile merchant family who is known for his flashy dress and his enthusiastic partying. After hearing a voice, which he believes to be that of God, Francis undergoes a religious transformation, forsakes partying, gives away his possessions, and even sells some of his father's textiles, giving away the money. His father responds by confining the youth to the house and beating him to bring him to his senses, but Francis is unrepentant.

Exasperated, the irate father takes Francis to court, which orders Francis to repay his father. In protest, Francis gives back everything his parents have given him, even the clothes off his back, and walks out of court naked. He forms a religious sect whose members sleep in abandoned churches, possess nothing, and are not above begging for their food. Never does he return to a normal social life.

For Francis (to whom we will shortly return) is religion beneficial or hazardous to mental health? For you and me, is religious devotion good or bad for mental health?

Our culture offers us two sharply contrasting answers. Pop Christianity assures us that with sufficient faith, prayer, and positive thinking we can get Jesus to lift our burdens, to exorcise the demonic within us, to heal our emotional agonies, even to bless us with prosperity. Television evangelists and religious paperbacks offer hopeful testimonies of how one can get God to give us happy homes, robust sex, inner peace, or liberation from depression. In Christian inspirational magazines one can find ads for things such as the "Christian weight loss plan," which promises results superior to those of non-Christian weight loss plans.

Opposing those who say that faith is the key to inner healing are those who say that religion erodes mental health or even that religion is a sickness—an "obsessional neurosis," said Freud. Religion is said to promote neurotic guilt, repression of sexual feelings, and suppression of negative emotions. George Albee, past president of the American Psychological Association, argues that religion also impedes efforts to relieve human misery by teaching that people deserve their fate; to believe that misfortune and suffering are divine judgments on sinners legitimates blaming the depressed, the miserable, and the angry for their feelings.

Who is right? Is religion more often beneficial or hazardous to mental health? Let's approach this question first scientifically, by looking at research on religion and mental health, and then theoretically, by reflecting on the likely emotional consequences of being a Christian disciple.

Religion and Mental Health: The Evidence

Are there any links between people's religiosity and their mental health? This simple question has no simple answer, because the answer depends on what we mean by religiosity (orthodoxy? church attendance? strength of religious feeling?) and what we mean by mental health (positive self-esteem? absence of mental illness? personal competence?).

Saying, "It all depends," leaves us wanting to know *what* the religion–mental health link depends upon. In their helpful book, *The Religious Experience*, social psychologists Daniel Batson and

Larry Ventis review the available research and provide some answers. If by "mental health" we mean acceptable social behavior and freedom from worry and guilt, then the dozen and a half available studies reveal no effect of being religious. If by mental health we mean feeling in control of one's life and expressing positive feelings about oneself, then the two dozen available studies point to a negative relationship: those who are more religious are less likely to say positive things about themselves. If by mental health we mean the absence of symptoms of mental illness, then the dozen available studies point to a positive relationship: religious people are less likely to be diagnosed as suffering a psychological disorder or to commit suicide.

Batson and Ventis report that the religion–mental health relationship also depends on people's religious motivation. Among religious people, those who are extrinsically motivated by the social and personal rewards of their religious affiliation tend to be plagued by worry, guilt, and the feeling that things are beyond control. Those for whom religion is an open-ended quest or who are *intrinsically* religious (those for whom religion is a master motive, an end in itself and not just a means to an end) tend to express better feelings about themselves.

A word of caution is in order: these studies merely establish a correlation between religion and mental health. It is a familiar lesson, but true: correlation does not indicate the direction of cause and effect. One's religion may indeed influence one's mental health. One's mental health may affect one's religion (some religious cults have been a haven for disturbed people). Or religiosity and mental health may be jointly influenced by underlying factors, such as one's social, economic, or educational status.

Religion and Mental Health: A Biblical Perspective

Will a real Christian ever act crazy? Indeed yes. If Christ's followers march to the sound of a different drummer in what they regard as a crazed world, they may, at times, seem a little crazy.

So it was with Francis, whom we today remember not as a lunatic but as the revered Saint Francis of Assisi, founder of the Franciscan order and a thirteenth century missionary and religious pioneer. Francis dared to be different, to renounce his family's materialism, to value higher things, and to suffer rejection for doing so.

And so it was with Jesus and some of his early followers. They knew negative emotions—righteous anger in response to injustice, anxiety when confronted by danger, grief in the face of death. They willingly experienced humiliation, even death, as the price for *not* adjusting to their culture. For the heroes of the Bible, good adjustment—thinking well of oneself and feeling positive about the world—was not the aim of life. As psychologist Raymond Paloutzian reminds us, "adjusting (or conforming) to a sick society may itself be a sick response."

It is ironic that popular religion should promise its followers serenity and success when the Bible itself depicts its people as so imperfect. The heroes of the faith experienced more tribulation than triumph. In the Old Testament, Noah becomes a drunken fool, David commits homicide out of lust, and Jacob is a blasphemous, polygamous, ungrateful cheat. Likewise, in the New Testament we find the afflicted Paul struggling constantly to resist what he ought not be doing and to do the good which he ought to be.

Moreover, one doubts that any of the disciples could have offered persuasive testimonies of "how I overcame anger, selfishness, and doubt." Peter loses his temper, is prejudiced against Gentiles, and denies Christ. After almost three years with Jesus, Andrew cannot conceive of a miracle with loaves and fish. The proud and prejudiced Nathaniel is skeptical that anything good could come out of Nazareth. Unless Jesus would "show us the Father," Philip refuses to believe that Jesus and God are one. The sons of Zebedee, James and John, crave the highest status positions for themselves in the coming kingdom. Thomas doubts Christ's resurrection and is skeptical of Jesus' promise to prepare a place in the Father's house. Simon the Zealot, Bartholomew, Matthew, and Jude can't manage so much as to stay awake during Jesus' agony before his betrayal.

The Bible makes no pretensions about the perfections of its people. Nor does it need to, for its hope rests not in the power of human faith but in the steadfast love of God.

As the experience of Job reminds us, God's people are not promised an earthly haven from misery. No matter how much faith we have, or how many faith healers we visit, our mortality rate will still be 100 percent. As the theologian Reinhold Niebuhr wrote, "It is easy to be tempted to the illusion that the child of God will be accorded special protection from the capricious forces of the natural world or special immunity from the vindictive passions of angry men. Any such faith is bound to suffer disillusionment." Better to root our faith in the hard truth than in temporarily comforting fantasies. If Christianity is untrue, then what honest person would want to believe it, however comforting it might be? And if it is true, even if it were not immediately comforting, what honest person would want to disbelieve it?

Among the capricious forces of the natural world are oppressive environments (in which, at times, it is perfectly natural to feel depressed), biochemical and neurological deficits (for which schizophrenia may be a natural outcome), and genetic predispositions to respond maladaptively to stressful circumstances. Faced with psychological disorders such as depression and schizophrenia, Christians had therefore best respond not with simplistic snap judgments (as Job's friends did in response to his misery) but with compassion and understanding. For some this may mean doing or supporting research. For others it means entering a helpful profession as a clinician, counselor, or social worker. For many more it simply means being loving, caring, and patient.

Although Christian faith does not promise escape from the stresses and woes of life, it can help us walk through the valley of deepest darkness. It does so first by offering us an identity—a knowledge of who we are, of our ultimate values, of our mission in life. The psychiatrist Carl Jung once declared that among his thousands of patients in the second half of life, "There has not been one whose problem in the last resort was not that of finding a religious outlook on life." More recent questionnaire studies confirm that adults who have a strong sense of purpose in life experience greater well-being, live with less dread of death, and are less likely to abuse alcohol and other drugs.

Second, religious communities offer social support in times of stress. Recent research indicates that people who are upheld by close relationships, such as in a close-knit religious fellowship, are less vulnerable to illness and premature death than are those who bear their stresses alone. When faced with a threat, caring friends can help us evaluate the problem, restore our self-esteem, reduce our anxiety, and confide our painful feelings—all of which can be good medicine.

Finally, religious experience has the potential to be therapeutic— at times by providing peak experiences of joy, peace, and enlightenment, but more often by reassuring us that, come what may, we are loved. Researchers have found that people's God-concepts are linked with their self-concepts: those who view God as stern and punitive tend to have low self-images; those who view God as loving and accepting tend to express higher self-esteem. And that leads us, in the next chapter, to a biblical key to psychological wholeness—the experience of grace.

For Further Reading

Batson, C. D. and W. L. Ventis. *The Religious Experience: A Social-Psychological Perspective*. New York: Oxford University Press, 1982.
A comprehensive but accessible summary of research on the nature and the consequences of religious experience.
Buchanan, D. *The Counseling of Jesus*. Downers Grove, Ill.: InterVarsity Press, 1985.
Offers Jesus' ways of dealing with anxious, angry, guilty, and hurting people as a model for our own dealings with such people.

Chapter 24

GRACE: GOD'S UNCONDITIONAL POSITIVE REGARD

Lord, I have given up my pride
    and turned away from my arrogance.
I am not concerned with great matters
    or with subjects too difficult for me.
Instead, I am content and at peace.
As a child lies quietly in its mother's arms,
    so my heart is quiet within me.

PSALM 131:1–2, TEV

In the *Interpreter's Dictionary of the Bible*, S. J. DeVries reduces the whole of Scripture to a single sentence: People find themselves "in sin and suffer its painful effects; God graciously offers salvation from it. This, in essence, is what the Bible is about."

Sin and salvation—what a contrast to other formulas for abundant life, especially to those that drive us to chase the rainbow of self-perfection. "You can do it," we are told. You can become assertive, wealthy, powerful, slim, sexy, happy, and healthy. And so we build our careers, our reputations, our bank accounts, our bodies. But self-perfection still eludes us. Our ambitions rise. New worries intrude: Will our grades be high enough to enable success? Will our next venture fail? Can we sustain the earning power our life style now demands? Will our children bring honor or embarrassment to ourselves and themselves? So we try harder and achieve more, and still the seductive voice whispers: "You can do it; you can be happy, secure, and peaceful, once your current ambitions are realized."

If, worse, we fail to achieve—if we get mediocre grades, are

unhappily married, poor, underemployed, overweight, or have rebellious children—we have ourselves to blame. Shame. We therefore seek to justify ourselves, in both our own and others' eyes.
Social psychologist Jerald Jellison, who has catalogued hundreds of
ways by which people try to justify themselves, laments that

the only way we have to cope with anxiety is to try to stay in the world's
good graces—by being "good," giving honest justifications whenever possible, and fabricating excuses when necessary. These lies, the criticism
of . . . others and the continual self-consciousness give rise to self-doubt
and uncertainty about our own worth. It is painful, and it is a pain familiar
to all of us.

One need not disparage grades, careers, finances, or families to
appreciate the skepticism of Ecclesiastes: "I have seen everything
that is done under the sun; and behold, all is vanity and a striving
after wind."

From this treadmill of self-perfection and self-justification the
Christian gospel promises liberation. In the Sermon on the Mount,
Jesus hints at the paradoxical way by which comfort, satisfaction,
mercy, peace, happiness, and visions of God are discovered: "Happy
are those who know they are spiritually poor; the Kingdom of
Heaven belongs to them!" We must begin by giving up our com-
pulsive strivings, by "losing ourselves," by relinquishing our vanity,
by becoming as unpretentious as a little child.

C. S. Lewis explains the irony of the sin-salvation sequence.
"Christian religion is, in the long run, a thing of unspeakable comfort. But it does not begin in comfort; it begins in [dismay], and it
is no use at all trying to go on to that comfort without first going
through that dismay." The first step toward wholeness and inner
peace is to acknowledge that self-interest and self-deception taint
every corner of our lives. The insights gleaned from psychological
research on illusory thinking and self-serving pride therefore have
deep Christian significance, for they reinforce the biblical view of
our human limits and our spiritual poverty.

Christians furthermore believe that a genuine experiencing of

God's grace liberates us from the need to define our self-worth in terms of achievements or prestige or material and physical well-being. Recall Pinocchio. Floundering in confusion about his self-worth, Pinocchio turns to his maker Gepetto and says, "Pappa, I am not sure who I am. But if I'm all right with you, then I guess I'm all right with me." In the life, death, and resurrection of Jesus, our Maker signals to us that we belong to God and that we are set right. Thus the Apostle Paul, surrendering his pretensions, could exult, "I no longer have a righteousness of my own, the kind that is gained by obeying the Law. I now have the righteousness that is given through faith in Christ."

"To give up one's pretensions is as blessed a relief as to get them gratified," noted the philosopher-psychologist William James, "and where disappointment is incessant and the struggle unending, this is what men will always do. The history of evangelical theology, with its conviction of sin, its self-despair, and its abandonment of salvation by works, is the deepest of possible examples." There is indeed tremendous relief in confessing our limits and our pride and in being known as we are. Having been forgiven and accepted, we gain release, a feeling of being given what formerly we were struggling to get: security, peace, love. Having cut the pretensions and encountered divine grace, we feel *more* not less value as persons, for our self-acceptance no longer depends exclusively upon our own virtue and achievement nor upon others' approval.

The feelings one can have in this encounter with God are like those we enjoy in a relationship with someone who, even after knowing our inmost thoughts, accepts us unconditionally. This is the delicious experience we enjoy in a good marriage or an intimate friendship, in which we no longer feel the need to justify and explain ourselves or to be on guard, in which we are free to be spontaneous without fear of losing the other's esteem. Such was the psalmist's experience: "Lord, I have given up my pride and turned away from my arrogance . . . I am content and at peace."

One way to comprehend this divine grace is to comprehend its human counterpart. Psychologist Carl Rogers contends that we nurture others' growth by being genuine (open and honest regarding our own feelings), by being empathic (listening with a reflective understanding of the others' feelings), and by being *accepting*—by offering *unconditional positive regard*. What better definition of grace than unconditional positive regard—knowing someone as he or she truly is and valuing the person nonetheless. From his half-century of experience in counseling, Rogers concludes, "As persons are accepted and prized, they tend to develop a more caring attitude toward themselves."

The practical effects of experiencing grace were demonstrated some years ago in an experiment by psychologist James Dittes. He temporarily led some Yale University students to feel either accepted or rejected by their peers. He then gave them a fabricated parable in biblical language and asked them to interpret it. Although the parable was actually incoherent (no two people agreed on what it meant), those who were feeling rejected tended unhesitatingly to assign it a meaning. (Doing so required ignoring some parts of the parable and distorting others to fit the supposed meaning.) The other students—those who had "experienced grace"—were less dogmatic and better able to acknowledge that to them the parable was ambiguous.

Such findings help us understand in human terms what grace can mean and how important it is that we be instruments of grace to those around us. It also helps us appreciate what it can mean to know that we are accepted just as we are, not just by other humans but by the very Lord of the universe. To those with a deep sense of God's love, this provides a more solid grounding for self-acceptance than do the ups and downs of others' approval. No longer do we need to establish our self-worth or to prove it to others. We need only to *accept* it.

To those who were burdened with the struggle to justify themselves by adherence to rules and laws, Jesus said, "Come to me, all you who labor and are heavy laden, and I will give you rest." This is the sense of unconditional positive regard, of ultimate acceptance—of amazing grace—that the apostle Paul proclaimed to the church in Rome:

I am convinced that there is nothing in death or life, in the realm of spirits or superhuman power, in the world as it is or the world as it shall be, in the forces of the universe, in heights or depths—nothing in all creation that can separate us from the love of God in Christ Jesus our Lord.

For Further Reading

Tournier, P. *Guilt and Grace: A Psychological Study.* New York: Harper & Row, 1962.
    A classic, helpful book by the famous Christian psychiatrist. "For a man crushed by the consciousness of his guilt, the Bible offers the certainty of pardon and grace." Tournier illustrates what this means both spiritually and psychologically.

Chapter 25

VALUES IN PSYCHOTHERAPY

Values influence every phase of psychotherapy, from theories of personality and pathology, the design of change methods, and the goals of
treatment, to the assessment of outcomes.
ALLEN E. BERGIN
"PSYCHOTHERAPY AND RELIGIOUS VALUES," 1985

One of our earlier themes—that those who view psychology through the eyes of faith should be alert to its hidden values
and assumptions—is especially pertinent to theories of personality
and psychotherapy. Some Christian psychologists therefore criticize
the philosophical assumptions that underlie the behaviorist psychology of B. F. Skinner—the assumptions that behavior is totally
determined, that people's thoughts merely accompany their behavior without causing it, and that principles of behavior control de-
rived from research with animals can be effectively and ethically
applied to the control of human behavior. Other Christian psychologists critique the assumptions of humanistic psychology—that
people are fundamentally good and not disposed to evil, that to be
true to oneself is the highest good, that self-analysis can reveal more
important truths than are revealed by scientific analysis.

But no Christian psychologist has provoked so much discussion
within psychology as has Brigham Young University psychologist
Allen Bergin, a respected psychotherapy researcher and coeditor of
the influential *Handbook of Psychotherapy and Behavior Change.*
In 1980, Bergin published an article in the American Psychological
Association's clinical psychology journal explaining his Judeo-
Christian values and contrasting them with the values assumed or
proclaimed by many other psychotherapists. The response was enormous. Several people submitted articles responding to Bergin, doz-

ens more have discussed Bergin's article in the six years since its publication, and a thousand more people have written him, offering their thoughts or requesting a reprint of the controversial paper.

And what were so many psychologists responding to? Bergin had nailed six theses to the door of clinical psychology:

Theses 1 and 2: Values Are an Inevitable and Pervasive Part of Psychotherapy

Every therapist has some idea of what mental health is, or at least of what is a better or a worse outcome to therapy. Should clients be encouraged to become more self-assertive or more other-focused? More accepting of things as they are or more ambitious? More open to uncommitted sex or more in control of their impulses? In answering such questions, said Bergin, every therapist is a moral agent. Even those who advocate so-called nondirective counseling have been observed to reward attitudes and expressions that they value—by smiling, nodding, repeating, or otherwise showing interest and approval.

Thesis 3: The Dominant Values in the Mental Health Profession Exclude and Clash with Religious Values

Some clinicians have pragmatic values, observed Bergin. They aim to reduce disturbances stemming from anxiety, depression, guilt, or deviant behavior. Other clinicians have humanistic values. They encourage self-exploration, self-fulfillment, and independence—values expressed in Carl Rogers's statement, "The only question which matters is, 'Am I living in a way which is deeply satisfying to me, and which truly expresses me?' " Given such values we can understand how self-denial could be labelled, as it was in a 1986 issue of *Psychotherapy*, a psychiatric syndrome—the "self-less syndrome," in which concern for meeting the needs of others supplants a "healthy self-centeredness."

Bergin wanted not so much to dispute such values as to reveal what they leave out—religious values such as the following:

Religious	Clinical-Humanistic
God is supreme. Humility, acceptance of (divine) authority, and obedience (to the will of God) are virtues.	Humans are supreme. The self is aggrandized. Autonomy and rejection of external authority are virtues.
Personal identity is eternal and derived from the divine. Relationship with God defines self-worth.	Identity is ephemeral and mortal. Relationships with others define self-worth.
Self-control in terms of absolute values. Strict morality. Universal ethics.	Self-expression in terms of relative values. Flexible morality. Situation ethics.
Love, affection, and self-transcendence are primary. Service and self-sacrifice are central to personal growth.	Personal needs and self-actualization are primary. Self-satisfaction is central to personal growth.
Committed to marriage, fidelity, and loyalty. Emphasis on procreation and family life as integrative factors.	Open marriage or no marriage. Emphasis on self-gratification or recreational sex without long-term responsibilities.
Personal responsibility for own harmful actions and changes in them. Acceptance of guilt, suffering, and contrition as keys to change. Restitution for harmful effects.	Others are responsible for our problems and changes. Minimizing guilt and relieving suffering before experiencing its meaning. Apology for harmful effects.
Forgiveness of others who cause distress (including parents) completes the therapeutic restoration of self.	Acceptance and expression of accusatory feelings are sufficient.

| Knowledge by faith and self-effort. Meaning and purpose derived from spiritual insight. Intellectual knowledge inseparable from the emotional and spiritual. Ecology of knowledge. | Knowledge by self-effort alone. Meaning and purpose derived from reason and intellect. Intellectual knowledge for itself. Isolation of the mind from the rest of life. |

Thesis 4: Psychologists' Values Differ From Those of Many of Their Clients

Surveys indicate that compared to people in general, and to psychotherapy clients in particular, psychologists are much more likely to be agnostic or atheistic, to approve of premarital sex, and in general to advocate liberal attitudes.

Thesis 5: It Would Be Honest and Ethical for Mental Health Workers to State Their Values Openly

Many psychologists, psychiatrists, and social workers mistakenly assume that what they are doing is value-free professionalism when in fact it is unavoidably value laden. "During my graduate and postdoctoral training," recalled Bergin,

I had the fortunate experience of working with several leaders in psychology, such as Albert Bandura, Carl Rogers, and Robert Sears. (Later, I had opportunities for substantial discussions with Joseph Wolpe, B. F. Skinner, and many others.) These were good experiences with great men for whom I continue to have deep respect and warmth; but I gradually found our views on values issues to be quite different. I had expected their work to be "objective" science, but it became clear that these leaders' research, theories, and techniques were implicit expressions of humanistic and naturalistic belief systems that dominated both psychology and American universities generally.

Subsequent to Bergin's essay, clinical researcher John Gartner asked all the clinical psychology professors in the American Psychological Association–approved doctoral programs to help him with a study of graduate student admissions. Each professor was asked to evaluate one mock application. Gartner's finding of antireligious prejudice would not surprise Bergin: applicants who identified themselves as an "evangelical fundamentalist Christian" were less likely to be rated as good prospects for admission than were otherwise identical applicants who said nothing about their religious interests. To declare oneself as a fervent, conservative Christian was, it seemed, to risk being perceived as a religious nut.

Thesis 6: The Effects of Different Value Systems Should Be Openly Tested and Evaluated

Bergin therefore offered his own values as testable hypotheses, such as:

religious communities that provide both a belief structure and loving, emotional support should manifest fewer psychological and physical disorders;

groups that endorse high standards of impulse control should have low rates of alcoholism, drug addiction, divorce, and emotional instability;

a stable marriage and family life pay psychological and social dividends.

Other hypotheses proposed the benefits of fidelity, of forgiveness, and of the acceptance of guilt and suffering.

In responding to Bergin, psychologists generally agreed that their values were important and should be acknowledged more openly. Some, however, complained that religious dogmatism and inhibitions are anything but healthy, and that clinical-humanistic values are more fully humanizing. Yes, replied Bergin, "religion is diverse" and not always benevolent, and, yes, as a religious person he strongly supported humanistic values of love, freedom of choice, and honesty. Nevertheless, his point remained, and still does: values that permeate psychotherapy should be openly acknowledged and tested; moreover, psychologists should consider the possibility that genuine spiritual-religious values may indeed have positive, health-promoting consequences.

For Further Reading

Kilpatrick, W. K. *Psychological Seduction: The Failure of Modern Psychology*. Nashville, Tenn.: Thomas Nelson, 1983.

As occasionally sarcastic but consistently provocative Christian critique of humanistic themes in personality and developmental psychology.

Vitz, P. C. *Psychology as Religion: The Cult of Self-Worship*. Grand Rapids, Mich.: Eerdmans, 1977.

This thought-provoking and widely quoted little book would have been more accurately entitled *Humanistic Psychology as Religion*; it critiques the "selfism" of four humanistic psychologists: Erich Fromm, Carl Rogers, Abraham Maslow, and Rollo May.

Wallach, M. A., and L. Wallach. *Psychology's Sanction for Selfishness: The Error of Egoism in Theory and Therapy*. New York: W. H. Freeman, 1983.

Two prominent psychologists offer a wide-ranging analysis of how psychology assumes and promotes selfishness.

Chapter 26

NICE PEOPLE AND EVILDOERS

Evil communications corrupt good manners.

1 CORINTHIANS 15:33, KJV

In his 1665 book of *Maxims*, the French writer and moralist Francois La Rochefoucauld noted, "Men's natures are like most houses—many sided; some aspects are pleasant and some not." So it is with human relations. No other species is capable of more brutal aggression against its own members, or of more compassionate giving of help. What other animals can feel such bitter prejudice toward another, or such profound love?

In this latter part of the twentieth century, our capacities for evil, armed with modern weaponry, have become more terrifying than ever. The Soviet and American nations each build their stocks of "defensive" nuclear weapons to protect themselves against domination by their "hostile" enemy. If such threats seem remote, many North Americans need look no further than their own neighborhoods to feel endangered. In the half-century since the 1930s the average person's vulnerability to robbery, assault, and murder increased severalfold. Women are now more often reported raped, children more often abused.

In much the same way that natural evils, such as earthquakes and tornadoes, can be studied by scientists, so, too, can behavioral scientists explore the workings of human evils. Some people attribute human evil to the dispositions of "rotten apples"—the freeloaders, thieves, rapists, and terrorists who threaten the rest of us nice folks. If evil is a personal abberation, then its remedy must also be personal. The way to alleviate unemployment is to put individuals through job training. The way to remedy emotional suffering is to put the individual in therapy. The way to deal with sin is to convert the individual.

But research clearly reveals that the human problem is also collective and that these individualistic remedies often deliver less than expected. True, evil emanates from the hearts of individuals. (Recall from Chapters 21 and 22 the phenomena of ever-escalating material greed and of self-serving pride.) Yet evil also accumulates into a power that transcends and corrupts even well-meaning individuals.

Laboratory experiments enable us to isolate some of the important features of social situations. By compressing social forces into a brief time period, we can see how these forces affect people. A number of such experiments have put well-intentioned people in an evil situation to see whether good or evil prevails. To a dismaying extent, evil pressures overwhelm good intentions, inducing people to conform to falsehoods or capitulate to cruelty. Nice guys often don't finish nice.

The clearest example is Stanley Milgram's obedience experiments. Faced with an imposing, close-at-hand commander, 65 percent of his adult male subjects fully obeyed instructions. On command, they would deliver what appeared to be traumatizing electric shocks to a screaming innocent victim in an adjacent room. These were regular people—a mix of blue-collar, white-collar, and professional men. They despised their task. Yet obedience took precedence over their own moral sense and the pleas of their victim.

As often happens in everyday life, the entrapment of these men was a subtle step-by-step process. At first, they had to perform only little cruel acts. Then the demands for evil were gradually escalated. Before long, the barely perceptible 15 volts had become a seemingly excruciating 450 volts. As Screwtape advised Wormwood, "The safest road to hell is the gradual one." Succumbing to just a little temptation begins to erode the conscience, making the next evil act possible.

In complex societies, a sequence of small evils can lead to vile evil. Nazi leaders were surprised at how easily they got German civil servants to handle the paperwork of the Holocaust. They were not killing Jews, of course. They were merely pushing paper. And so it is in our daily lives: the drift toward evil usually comes in small steps, without any conscious decision to do evil. C. S. Lewis illustrated the subtlety of this corrupting process:

Over a drink or a cup of coffee, disguised as a triviality and sandwiched between two jokes, from the lips of a man, or woman, whom you have recently been getting to know rather better and whom you hope to know better still—just at the moment when you are most anxious not to appear crude, or naif or a prig—the hint will come. It will be the hint of something which is not quite in accordance with the technical rules of fair play: something which the public, the ignorant, romantic public, would never understand: something which even the outsiders in your own profession are apt to make a fuss about: but something, says your new friend, which "we"—and at the word "we" you try not to blush for mere pleasure—something "we always do." And you will be drawn in, if you are drawn in, not by desire for gain or ease, but simply because at that moment, when the cup was so near your lips, you cannot bear to be thrust back again into the cold outer world. It would be so terrible to see the other man's face—that genial, confidential, delightfully sophisticated face—turn suddenly cold and contemptuous, to know that you had been tried for the Inner Ring and rejected. And then, if you are drawn in, next week it will be something a little further from the rules, and next year something further still, but all in the jolliest, friendliest spirit. It may end in a crash, a scandal, and penal servitude: it may end in millions, a peerage, and giving prizes at your old school. But you will be a scoundrel.

Procrastination involves a similar unintended drift toward harm. A student knows weeks ahead the deadline for a term paper. Each diversion from work on the paper—a video game here, a TV program there—seems harmless enough. Yet the student gradually veers toward not doing the paper (or at least not doing it well) without consciously deciding not to do it.

That nice people can be corrupted by evil situations is evident even in experiments that involve only minimal coercive social pressure. Numerous studies indicate the powerful effects of occupying a social role. Although a new role may at first feel artificial—we can feel that we are "playing" it—the sense of phoniness soon tapers off. We absorb the role into our personalities and attitudes. Participating in destructive roles can therefore corrupt a person. Soldiers, for example, almost unavoidably develop degrading images of their enemy.

Philip Zimbardo's prison simulation illustrates this vicious process. Among a volunteer group of decent young California men, some were randomly designated guards. They were given uniforms, billy clubs, and whistles. And they were instructed to enforce certain rules. The remainder became prisoners, locked in barren cells and forced to wear humiliating outfits.

After a day or two of role playing, the guards and prisoners, and even the experimenters, became caught up in the situation. The guards devised cruel and degrading routines. The prisoners either broke down, rebelled, or became passive. And the experimenters worked overtime to maintain prison security.

This demonstration's prime lesson concerns more than the dehumanizing effects of guard-prisoner role relations. The steel bars symbolize as well the way in which destructive role relations can affect rich and poor, white and black, husband and wife, employer and employee, teacher and student. Evil behavior is sometimes structured into the very roles we are forced to play.

Another potentially corrupting social force is competition for scarce goods and resources. Muzafer Sherif and his colleagues demonstrated the potentially destructive effects of competition when they offered a summer camp experience to a large number of healthy, unacquainted eleven- and twelve-year-old midwestern boys. They began by ascertaining what seemed to be the necessary conditions for hostility between groups. They then introduced these apparent essentials to see if the expected prejudice and violence would actually occur.

The boys were randomly divided into two groups. For a time, they lived separately with their group and cooperated on various tasks, thus establishing their group identity. The groups were then brought together for a series of competitive activities, with prizes only to the winning team.

Before long, the two groups of ordinary boys were calling each other names, making derogatory posters, hurling food, and attacking each other's campsites. It was warfare on a twelve-year-old scale, a living version of William Golding's *Lord of the Flies*—despite there being no cultural, physical, or economic differences between the groups. As in the other experiments, the evil outcomes reflected the power of an evil situation.

Dozens of other experiments reveal that groups often display in more striking form the individual tendencies of their members. It's true of our better tendencies—fellowship may fan the flames of our compassion, faith, and hope. And it's true of our baser tendencies—group interaction may also exacerbate meanness and self-righteousness. Self-serving perceptions may therefore mushroom into collective pride; racists, sexists, and nationalists all come to perceive that their group—their race, sex, or country—is superior to other groups. The Soviet leaders convince one another of their moral superiority, reasoning that their arms buildup is simply a self-protective response to America's evil power; the American leaders similarly convince themselves of their own moral superiority vis-à-vis the Soviet "evil empire." Thus Lewis Thomas's dismal conclusion:

For total greed, rapacity, heartlessness, and irresponsibility there is nothing to match a nation. Nations, by law, are solitary, self-centered, withdrawn into themselves. There is no such thing as affection between nations, and certainly no nation ever loved another. They bawl insults from their doorsteps, defecate into whole oceans, snatch all the food, survive by detestation, take joy in the bad luck of others, celebrate the death of others, live for the death of others.

These social-psychological examples of collective evil parallel biblical teachings about evil. Evil's external sources are represented in the story of the Fall as Eve is seduced by an external demonic force. The social character of sin is further evident in the corporate personality of Israel, in which whole families and whole cities are sometimes condemned for their wickedness. New Testament images of "principalities and powers" reinforce this notion that evil transcends individuals.

Precisely because sin has a collective aspect, we must also make a collective response to it. Judaism and Christianity are distinguished not by the mystical visions of spiritual isolates but by a community life. Since we are not self-sufficient, we stand in need of the church's corporate fellowship. Only in that context can we adequately struggle with the evil within and about us.

It is the whole believing people, not isolated believers, that is the body of Christ. To say the church is Christ's body reminds us that together we can admonish one another. Together we can enable each other to minister. And together we can contest evil in ways that we as individuals never could, by challenging and reforming not only the individual rotten apples, but also the social systems and situations that can make the barrel go bad. To repeat, evil is collective as well as personal and so requires a collective as well as a personal response.

For Further Reading

Gilkey, L. *Shantung Compound.* New York: Harper & Row, 1966.
> Theologian Gilkey tells a gripping true story about the corrupting power of an evil situation—a World War II Japanese internment camp into which were herded 1800 "nice" doctors, missionaries, and businesspeople. Must reading for anyone interested in looking beneath the veneer of human manners.

Menninger, K. *Whatever Became of Sin?* New York: Hawthorn Books, 1973.
> The dean of American psychiatry awakened readers with this unexpected book, reminding us of what had become a taboo subject—our human propensity to evil.

Stringfellow, W. *An Ethic for Christians and Other Aliens in a Strange Land.* Waco, Tex.: Word Books, 1973.
> A troubling but prophetic book that prepares us to recognize and address evil and its subtle manifestations in the world.

Chapter 27

BEHAVIOR AND ATTITUDES—ACTION AND FAITH

Only he who believes is obedient, and only he who is obedient believes. . . . Jesus says: "First obey, perform the external work, renounce your attachments, give up the obstacles which separate you from the will of God." Do not say you have not got faith. You will not have it so long as you persist in disobedience and refuse to take the first step.

DIETRICH BONHOEFFER,
THE COST OF DISCIPLESHIP, 1937

P eople generally assume that our beliefs and attitudes determine our actions. So if we want to change the way people act, their hearts and minds had better be changed. This assumption lies behind most of our teaching, preaching, counseling, and child rearing. And to some extent it's true: *behavior follows attitudes.* But if social psychology has taught us anything during the last thirty years it is that the reverse is also true: we are as likely to act ourselves into a way of thinking as to think ourselves into action.

Various streams of evidence have converged to establish this *attitudes follow behavior* principle:

As we saw in the preceeding chapter, evil acts shape the self. People induced to harm an innocent victim typically come to disparage their victim. Those induced to speak or write statements about which they have misgivings will often come to accept their little lies. Saying becomes believing.

Moral action effects the actor, too. Children who resist a temptation tend to internalize their conscientious behavior. Helping someone typically increases liking for the person helped. Those who teach a moral norm to others subsequently follow the moral code better themselves. Generalizing the principle, it would seem that one antidote for the corrupting effects of evil action is repentant action. Act as if you love your neighbor—without worrying whether you really do—and before long you will feel a greater liking for the person.

Racial attitudes have followed racial behavior. Prior to desegregation in the United States it was often said that you can't legislate racial attitudes—you must wait for the heart to change first. However, after the initiation of desegregation white racial attitudes became noticeably less prejudiced. Moreover, as different regions of the country have come to act more alike, they have also come to think more alike.

Political socialization techniques have effectively employed the principle, from the daily flag salute to the Nazi "Heil Hitler" greeting. Historian Richard Grunberger reports, "Many Germans experienced discomfort at the contradiction between their words ["Heil Hitler"] and their feelings. Prevented from saying what they believed they tried to establish their psychic equilibrium by consciously making themselves believe what they said."

Many modern therapy techniques make a more constructive use of action. Behavior therapy, assertion training, and rational-emotive therapy all enable their clients to rehearse and practice more productive behavior. We can all learn a practical lesson here. Like Moses, Jonah, and other biblical heroes, we often do not feel like doing what we know we ought. The remedy is to get up and act anyway—to put the piece of paper in the typewriter and force ourselves to begin that essay or letter, to go to the phone and dial that number, to confront or share with that person, to turn off the TV and begin studying for that exam. When we do so we often find that our forced behavior begins to gain momentum as a real interest in our subject takes hold. Our feelings are hard to control, but we can control our behavior and by doing so indirectly influence our feelings.

To be sure, the attitudes-follow-behavior principle is more potent in some situations than others—especially in those where people feel some choice and responsibility for their behavior rather than attributing it to coercion. Nevertheless, it is now a fundamental rule of social psychology that *behavior and attitude generate one another in an endless spiral, like chicken and egg.*

This principle affirms the biblical understanding of action and faith, or what Bonhoeffer called obedience and belief. Depending on where we break into this spiraling chain, we will see faith as a source of action or as a consequence. Action and faith, like action and attitude, feed one another.

Much as conventional wisdom has insisted that our attitudes determine our behavior, Christian thinking has usually emphasized faith as the source of action. Faith, we believe, is the beginning rather than the end of religious development. The experience of being "called" demonstrates how faith can precede action in the lives of the faithful. Elijah is overwhelmed by the Holy as he huddles in a cave. Paul is touched by the Almighty on the Damascus Road. Ezekiel, Isaiah, Jeremiah and Amos are likewise invaded by the Word, which then explodes in their active response to the call. In each case, an encounter with God provoked a new state of consciousness, which was then acted upon.

This dynamic potential of faith is already a central tenet of Christian thought. For the sake of balance, we should also appreciate the complementary proposition: *Faith is a consequence of action.* Throughout the Old and New Testaments we are told that full knowledge of God comes through actively doing the Word. Faith is nurtured by obedience.

Reinhold Niebuhr and others have called attention to the contrast in assumptions between the platonic thought that permeates our Western culture today and biblical thought. Plato presumed that we come to know truth by reason and quiet reflection. This view, translated into Christian terms, equates faith with cerebral activity—orthodox doctrinal propositions.

The contrasting biblical view assumes that reality is known through obedient commitment. As O. A. Piper has written in the *Interpreter's Dictionary of the Bible,* "This feature, more than any other, brings out the wide gulf which separates the Hebraic from the Greek view of knowledge. In the latter, knowledge itself is purely theoretical . . . whereas in the Old Testament the person who does not act in accordance with what God has done or plans to do has but a fragmentary knowledge." For example, the Hebrew word for *know* is generally used as a verb, as something one does. To know love, we must not only know about love but we must act lovingly. Likewise, to *hear* the word of God means not only to listen but also to obey.

We read in the New Testament that by loving action a person knows God, for "he who does what is true comes to the light." Jesus declared that whoever would do the will of God would know God, that he would come and dwell within those who heed what he said, and that we would find ourselves not by passive contemplation but by losing ourselves as we take up the cross. The wise man, the one who built his house on rock, differed from the foolish man in that he acted on God's Word. Merely saying "Lord, Lord" does not qualify us as disciples; discipleship means doing the will of the Father. Over and over again, the Bible teaches that the gospel power can only be known by living it.

Our theological understanding of faith is informed by this biblical view of knowledge. Faith grows as we act on what little faith we have. Just as experimental subjects become more deeply committed to something for which they have suffered and witnessed, so also do we grow in faith as we act it out. Faith "is born of obedience," said John Calvin. "The proof of Christianity really consists in 'following,' " declared Søren Kierkegaard. Karl Barth agreed: "Only the doer of the Word is its real hearer." Pascal is even more plainspoken: To attain faith, "Follow the way by which [the committed] began; by acting as if they believed, taking the holy water, having masses said, etc. Even this will naturally make you believe. . . ." C. S. Lewis echoed Pascal's sentiments:

Believe in God and you will have to face hours when it seems *obvious* that this material world is the only reality: disbelieve in Him and you must face hours when this material world seems to shout at you that it is *not* all. No conviction, religious or irreligious, will, of itself, end once and for all [these doubts] in the soul. Only the practice of Faith resulting in the habit of Faith will gradually do that. . . .

People . . . are told they ought to love God. They cannot find any such feeling in themselves. What are they to do? The answer is . . . act as if you did. Do not sit around trying to manufacture feelings. Ask yourself, "If I were sure that I loved God, what would I do?" When you have found the answer, go and do it.

How can we apply the faith-follows-action principle to leading a church, to planning worship, and to nurturing personal faith? First, a top priority for churches must be to make their members active participants, not mere spectators. Many dynamic religious movements today—ranging from sects like the Jehovah's Witnesses, Mormons, and the Unification Church to charismatics and discipleship-centered communities—share an insistence that all on board be members of the crew. That is easier said than done, but it does provide a criterion by which to evaluate procedures for admitting and maintaining members. As a local church makes decisions and administers its program, it should constantly be asking, will this activate our people and make priests of our believers? If research on persuasion is any indication, this will best be accomplishehd by direct, personal calls to committed action, not merely by mass appeals and announcements.

In worship, too, people should be engaged as active participants, not as mere spectators of religious theater. Research indicates that passively received spoken words have surprisingly little impact on listeners. Changes in attitude resulting from spoken persuasion are less likely to endure and influence subsequent behavior than attitude changes emerging from active experience. What's needed is to have listeners rehearse and act on what they hear. When the people sing responses, write their own confessions, contribute prayer, read Scripture responsively, take notes on the sermon, utter exclamations, bring their offerings forward, pass the peace, make the sign of the cross, or sit, stand, and kneel—acts that viewers of the electronic church do not perform—they are making their worship their own.

The principle has its limits, of course. We can become so preoccupied with doing things that we no longer have time to quietly receive God's Word of grace and direction for our lives. Like the Pharisees, we can substitute our deeds for God's act or think that any kind of action will do. To say that action nurtures growth in faith is not to tell the whole story of faith. But it does tell part of the story.

The action-attitude interplay can also inform Christian education and Christian nurture. Researchers have found that the attitudes most likely to affect our actions are those that we form by experience. We might therefore consider new methods of encouraging faith. Few Christian families appreciate and reap the benefits of family worship. Old Testament family practices helped people remember the mighty acts of God. When today's Jewish family celebrates the Passover by eating special foods, reading prayers, and singing psalms, all of which symbolize their historical experience, they are helped to renew the roots of deep convictions and feelings. Among Christians, family celebrations are becoming more common during Advent. With a boost from the church, home-based activity could be extended to celebrate all the great themes of the church year.

Although church and family ritual may sometimes degenerate into a superficial religious exercise, few of us appreciate the extent to which the natural ritual of our own personal histories has shaped who we are. Many of the things we did without question in childhood have long since become an enduring part of our self-identities. Indeed, because we have internalized our own rituals, we find it difficult to recognize them as rituals; but it is easy to recognize other people's rituals.

The overarching objective on which all these points converge is this: we need to create opportunities for people to enact their convictions, thereby confirming and strengthening their Christian identity. Biblical and psychological perspectives link arms in reminding us that faith is like love. If we hoard it, it will shrivel. If we use it, exercise it, and express it, we will have it more abundantly.

For Further Reading

Bonhoeffer, D. "The Call to Discipleship." Chapter 2 in *The Cost of Discipleship*. New York: Macmillan, 1963.
   A beautiful exposition of the interplay between action and faith— or, as Bonhoeffer calls it, between obedience and belief.

Chapter 28

THE FRUIT OF THE SPIRIT

Now the works of the flesh are plain: fornication, impurity, licentious-
ness, idolatry, sorcery, enmity, strife, jealousy, anger, selfishness, dissen-
sion, party spirit, envy, drunkenness, carousing, and the like. I warn you,
as I warned you before, that those who do such things shall not inherit
the kingdom of God. But the fruit of the spirit is love, joy, peace, patience, kindness, goodness, faithfulness, gentleness, self-control; against
such there is no law.

GALATIANS 5:19–23, RSV

For many people the most troubling thing about Christianity is some deep philosophical problem. For reasons described
in Chapter 6, we can live peaceably with impenetrable philosophical puzzles. What troubles us is something more concrete: the
behavior of many people who count themselves as disciples of the
one who taught, "You shall love your neighbor as yourself," and
even "Love your enemies and pray for those who persecute you."

Every religion is plagued by those who exploit it for their own
purposes or embarrass and defame it by their behavior. Thus every
skeptic, and indeed every believer, can point to those whose lives
suggest that religion is a sham—those who profess the love of Christ
and practice hate, who preach honesty and fail to report all their
income, who proclaim the unity of the church and attack people
whose doctrinal views differ from their own, who promote selfless-
ness and are vain to the core, who pretend concern and couldn't
care less. As this book goes to press, the newsmagazine covers picture a television evangelist who admits an adulterous affair and,
worse, to having used his million-dollar-plus income for luxurious
mansions with gold-plated fixtures. By the time you read this book,
the phenomenon will have been reenacted with fresh examples. As

Madeline L'Engle lamented, "Christians have given Christianity a bad name."

But then again, every religion can also point to those who exemplify its aspirations and whose lives inspire others to join them in their pilgrimage. Christianity, too, can point to its Martin Luther Kings and Mother Theresas, its Albert Schweitzers and Desmond Tutus, its contribution to the spread of hospitals and universities and to the abolition of slavery.

The vivid examples—the worst and the best—capture our attention but do not decide the issue: do self-professed Christians more than others tend to display the fruit of the Spirit, or their opposites? Is the Christian religion a source more of compassion or intolerance? The extremes—the church-going civil rights activists and the church-going Ku Klux Klan—cancel each other out. So it remains for dispassionate research with ordinary people to help us decide the issue.

In some respects, the links between religion and moral behavior are clear. Self-described Christians in the Western world engage in much less sexual promiscuity, much less drug and alcohol abuse, and much less violent crime and delinquent behavior. Because religion-morality links are correlational, the direction of cause and effect is sometimes ambiguous. Nevertheless, such findings hint that when the church is clear and forceful in its ethical prescriptions, it may be influential.

Many additional—and more perplexing—studies have accumulated on the links between religion and altruism and between religion and prejudice. Most studies of altruistic behavior have observed people's willingness to help in minor emergencies—to mail an addressed lost letter found on the sidewalk, to call the garage for a stranded motorist who just spent her only quarter on a wrong number call, to aid someone in an adjacent room who was heard to fall off a ladder. Researcher Daniel Batson has reviewed the available research, and his conclusion would not have pleased the apostle Paul:

There is no evidence that [highly religious people] are any more likely than the less religious to help someone in need. The more religious may *see* themselves as more helpful and caring; they may even be seen this way by others. But when it comes to action, there is no evidence that they are more helpful.

At first blush, there is even more cause for alarm in the findings of the religion-prejudice studies. American church members have tended to be *more* racially prejudiced than nonmembers, and those professing traditional Christian beliefs have expressed *more* racial prejudice than those with less traditional beliefs. Perhaps it shouldn't shock us, for throughout history religion has provided convenient excuses—indeed powerful justifications—for all sorts of cruelty. For the horrors of military crusades. For the dehumanization of slavery and apartheid. For the subordination of women. The beautiful medieval town of St. Andrews, where these words are being written, was the ecclesiastical center of early Protestantism in Scotland. In the year 1643 alone—the midpoint of a 150-year reign of terror in St. Andrews and its environs—forty terrified women were judged by church elders to be witches and consigned to torture and death. These women and the St. Andrews martyrs who preceded them at the time of the Reformation remind us that behind religious fanaticism evil sometimes lurks. Jesus therefore reserved some of his strongest condemnation for the self-righteous religious folk of his day. From his time to ours, "not everyone who says . . . 'Lord, Lord,' " speaks for God. As Pascal lamented, "Men never do evil so completely and cheerfully as when they do it from religious conviction."

These things are true, but they are not the whole truth about the consequences of religion. Studies of more long-term altruism, though fewer in number, hint that self-described Christians tend to be more self-giving—more likely to give away significant amounts of money and more likely to have devoted many hours during the preceding year in volunteer activities. For example, among the 12 percent of Americans whom George Gallup in 1984 classified as "highly spriritually committed," 46 percent said they were presently working among the poor, the infirm, or the elderly—many more than among those less committed.

In the realm of prejudice, religion's role seems paradoxical. As psychologist Gordon Allport noted three decades ago, "It makes prejudice and it unmakes prejudice." The unmaking of prejudice is suggested first by studies of church members; in nearly every one of more than two dozen studies, faithful church attenders exhibited less prejudice than irregular attenders. Second, those for whom religion is an end in itself (who agree, for example, with the statement "My religious beliefs are what really lie behind my whole approach to life") typically express less racial prejudice than those for whom religion is more a means to other ends (for example, who agree that "A primary reason for my interest in religion is that my church is a congenial social activity"). Third, ministers and priests—who presumably are more religiously committed and motivated than most people—have also generally been more supportive of civil rights efforts than have their own laypeople. So it seems that among the churched, the devout exhibit less prejudice and deeper feelings of human brotherhood and sisterhood than the nominally religious, who are somewhat more likely to rationalize prejudice with the aid of religion. "We have just enough religion to make us hate, but not enough to make us love one another," lamented the eighteenth-century satirist Jonathan Swift.

What, then, might be our response to bigots and their bigotry? *Hate the sin and love the sinner*. Hate the bigotry and love the bigot. Be intolerant of intolerance, despise lovelessness, detest injustice, and remember: "The fruit of the spirit is love, joy, peace, patience, kindness, goodness, faithfulness, gentleness, self-control."

Second, take heart from those heroes of the faith who exemplify such fruit. If we are most troubled by the smallmindedness of those whose lives seem to deny the good news message of love, peace, and reconciliation, we are also most encouraged by those whose lives witness to the power of deep faith.

One such person was the serene, unpretentious, soft-spoken Eric Liddell, who, thanks to the Oscar-winning movie *Chariots of Fire* is known to the world as a man who was exceptionally committed to his principles. Rather than run on Sunday, he gave up his chance for a likely Olympic gold medal in the hundred meters, suffered the insult of being called a traitor to his country for doing so, and then astonished everyone by instead running and winning the four hundred–meter race in world record time. Although Liddell returned home a national hero, his greater heroism began where the movie ends. Shunning fame, fortune, and the next Olympic games, he slipped out of the limelight to become a missionary to China, where he taught chemistry and English and later worked in rugged conditions among rural, peasant people amidst suffering and death triggered by Japan's invasion of China during the late 1930s.

By all accounts, Liddell unfailingly radiated good humor and kindness, and because of his smiling good nature was often a peacemaker in times of conflict among the peasants and between them and their invaders. Nor was he one to pass by on the other side of the road when someone was suffering or in need of a daring rescue effort. When, shortly before Japan entered World War II, his pregnant wife and two daughters left China for the safety of home, Liddell stayed behind to minister, and in 1943 he was rounded up along with 1800 other foreigners into a Japanese internment camp in the Shantung Province of North China. In Langdon Gilkey's 1966 book *The Shantung Compound: The Story of Men and Women Under Pressure*, Gilkey recalls the conflicts and selfishness that predominated among this assortment of businesspeople, missionaries, doctors, professors, junkies, and prostitutes, all crammed into a former mission station no longer than two football fields and not as wide. Subjected to privation but not torture, malnutrition but not starvation, the "fundamental bent of the total self in all of us was inward, toward our own welfare," observed Gilkey. "And so immersed were we in it that we hardly seemed able to see this in ourselves."

During his two years in the camp, Eric Liddell emerged as its "most outstanding personality," as another book on the Shantung

Compound later described him—the one "with a permanent smile." It was he who organized games and worship, taught science to the children, and cared for people of every sort. One Russian prostitute, for whom he put up some shelves, said he was the only man who did anything for her without wanting to be repaid. Gilkey's stark account of self-righteousness and self-centeredness within the camp is broken by this ray of light:

It is rare indeed that a person has the good fortune to meet a saint, but he came as close to it as anyone I have ever known. Often in an evening of that last year I (headed for some pleasant rendezvous with my girlfriend) would pass the games room and peer in to see what the missionaries had cooking for the teen-agers. As often as not Eric . . . would be bent over a chessboard or a model boat, or directing some sort of square dance— absorbed, warm and interested, pouring all of himself into this effort to capture the minds and imaginations of those penned-up youths. If anyone could have done it, he could. A track man, he had won the 440 in the Olympics for England [Britain, actually], and then had come to China as a missionary. In camp he was in his middle forties, lithe and springy of step and, above all, overflowing with good humor and love of life. He was aided by others, to be sure. But it was Eric's enthusiasm and charm that carried the day with the whole effort.

In all the accounts, Liddell emerges as a sort of contemporary Christ figure, a man whose life was empowered by the hour of prayer, Bible reading, and meditation with which he began each day before the others were awake; a man who according to his closest comrade was "literally, God-controlled, in his thought, judgment, actions, words"; a man who befriended despised prostitutes and business people and bridged the gulf between them and the missionaries; a man who could be seen carrying coal for an old person; a man who offered to sell his Olympic gold watch to buy more sports gear for the children; a man who, weakened by privation and hunger, began quietly to suffer headaches and discouragement, the early signs of a brain disease that, before many even realized he was seriously ill, took his life just months before the camp's liberation. As he lay in the arms of a friend, nurse Annie

Buchan, he spoke his last words, "Annie, it's complete surrender," then convulsed, vomited all over her, lapsed into a coma, and died within hours. Scotland mourned its lost hero, the one in whom the fruits of the Spirit were so clearly displayed: love, joy, peace, patience kindness, goodness, faithfulness, gentleness, self-control.

Again, what troubles us most about Christianity is not so much the philosophical puzzles as the seemingly un-Christian behavior of so many of us Christians. But then again, what inspires us most is not so much the theological genius of great Christian thinkers as it is the witness of unpretentious lives empowered by faith—of the Eric Liddells and of those individuals known to each one of us who in their own ways testify to the peace and love that can flow from lives touched by grace. It is people such as these who encourage us in times of uncertainty, discouragement, and self-pity to keep on—to fight the fight, to finish the race, to keep the faith.

For Further Reading

Batson, C. D., and W. L. Ventis. *The Religious Experience: A Social Psychological Perspective.* New York: Oxford University Press, 1982. Includes a 150-page review of research on the "consequences of the religious experience" for personal freedom, mental health, and brotherly and sisterly love.

Chapter 29

MORE SENSE AND NONSENSE

At the heart of science is an essential tension between two seemingly contradictory attitudes—an openness to new ideas, no matter how bizarre or counterintuitive they may be, and the most ruthless skeptical scrutiny of all ideas, old and new. This is how deep truths are winnowed from deep nonsense.

CARL SAGAN,
"THE FINE ART OF BALONEY DETECTION," 1987

It puzzles and perturbs us that so many people know so little of what research psychologists take to be the plainest facts of behavior and mental processes, while believing so many things that seem plainly wrong. But, you may ask, how does one winnow sense from nonsense? Who's to say whether faith healing is medically potent? whether the quality or trauma of one's birthing experience is life forming? whether hypnotized people can experience accurate memories of their childhoods or even of a previous life?

Carl Sagan advises disciplined analysis of all such questions— an empirical attitude like that occasionally found in Scriptures. "If a prophet speaks in the name of the Lord and what he says does not come true," cautioned Moses, "then it is not the Lord's message." Some claims—that God exists, that death is not final—are beyond verification. With other claims, the proof is in the pudding. Magician James Randi illustrates the empirical approach in conversing with those who claim to see auras surrounding people's bodies:

RANDI: Do you see an aura around my head?

AURA-SEER: Yes indeed.

RANDI: Can you still see the aura if I put this magazine in front of my face?

AURA-SEER: Of course.

RANDI: Then if I were to step behind a wall barely taller than I am, you could determine my location from the aura visible above my head, right?

Randi reports that no aura-seer has yet agreed to take this simple test.

When subjected to skeptical scrutiny, astonishing claims sometimes are supported. During the eighteenth century, scientists scoffed at the notion that meteorites had extraterrestrial origins. When two Yale scientists dared deviate from the conventional opinion, Thomas Jefferson jeered, "Gentlemen, I would rather believe that those two Yankee Professors would lie than to believe that stones fell from heaven." Sometimes, skeptical scrutiny refutes the skeptics.

More often it relegates unconventional claims to the mountain of amusing fables, atop long-forgotten claims of perpetual motion machines, cancer cures, or personality assessments through skull readings. Sifting reality from fantasy therefore requires being skeptical but not cynical, open but not gullible. Such is fundamentally a Christian attitude, one that allows the data of God's creation to restrain our imagination.

In this book we have suggested several examples of how skeptical scrutiny has exposed what now seems like nonsense—regarding claims of ESP, of subliminal backmasking effects in rock albums, of the accuracy of unchecked intuitive judgments, of the origins of sexual orientation. We also have seen how skeptical scrutiny is revealing deep insights into sensation, consciousness, memory, and attitudes. To cap our review, consider a dozen additional claims that look more and more like nonsense, and another dozen that look more and more like actualities.

Myths of Pop Psychology

1. *The brain has precise control centers for complex psychological states.* Scientific abilities are said to be "located in" the left hem-

isphere, artistic abilities in the right hemisphere. Aggression is said to be located in a neural structure called the amygdala, sex in the hypothalamus. As with many myths, there are elements of truth here. Each of these regions do perform certain essential functions for these psychological states. Destroy the region and the behavior may be lost, much as taking out the carburetor precludes driving a car. But the truth is that complex behaviors, from science to sex, require coordinated neural activity in many areas of the brain. The brain is an intricate whole system.

2. *The first minutes and hours after birth are critical for mother-infant bonding.* Mothers and newborns given prolonged physical contact supposedly form closer attachments, with the bonded infants deriving lasting physical, mental, and social benefits. But developmental psychologists who have scrutinized the available research are unconvinced; the benefits of bonding seem at best minimal and temporary. Moreover, given the underdeveloped state of the infant's brain and the impermanence of infant learning experiences, it becomes difficult to imagine how such a minuscule slice of one's early life history could possibly have enduring consequences. Mothers who have had cesarean deliveries and parents of adopted children can relax: you and your child have not missed out on something terribly important.

3. *As part of their passage to middle adulthood, men in their early forties undergo a traumatic midlife crisis.* As it becomes apparent that one's dreams are illusory, that work and family are not ultimately satisfying, one purportedly enters an agonizing period of turmoil, despair, and possibly a search for new meanings and relationships. Here, too, there may be a germ of truth—at least for some of those who, like the small sample of subjects in two famous studies, are high achieving, career-oriented, American males. But the truth is that studies of broader samples reveal that job and marital dissatisfaction, divorce, anxiety, and suicide do not surge during the early forties. More important than age per se are childbearing and nest-emptying, relocation, divorce, retirement, and other sig-

nificant life experiences that, whenever they occur, mark a transition to a new life stage.

4. *After age 65, people become more susceptible to short-term illnesses, less satisfied with life, and more fearful of death.* It's true that some abilities decline in later life. Our recall of newly learned information slows; our visual acuity, hearing, muscle strength, and stamina diminish; our capacity for fast and fluid thinking decreases. But in older age people generally retain their accumulated knowledge and wisdom, are less bothered by flu and other temporary illnesses, express relatively high levels of life and marital satisfaction, and, especially if they look back on life with a sense of fulfillment, calmly accept their own mortality.

5. *Myths of sleep: some people dream every night, others seldom dream; sleepwalkers are acting out their dreams, sleeptalkers are verbalizing their dreams; one can learn while sleeping.* Sleep researchers report, to the contrary, that during a night's sleep everyone has dreaming periods (about every ninety minutes), during which the muscles of our internally aroused body are, except for an occasional twitch, immobilized; sleepwalking and sleeptalking occur during deep sleep. In learning, as elsewhere, there is no free lunch; studying *before* sleep can be effective, but students do not process or retain taped information played while they sleep.

6. *Our past experiences are all recorded in our brains; with brain stimulation or with hypnosis, one can uncover and relive long-buried memories.* Far from having memories like video recorders, we construct our memories as we retrieve them, based partly on shreds of stored information, partly on what we now assume. The occasional "memory flashbacks" reported by stimulated patients during brain surgery actually have a dreamlike quality, often involving locations the person has never visited. Hypnotized subjects similarly recall childhood events with astonishingly vivid detail and astonishingly little accuracy; they typically act as adults imagine, say, a four-year old would talk and act. Likewise, those regressed to their "past lives" enact their presumptions of how people of the time would have

acted; when asked on separate occasions to recall prior lives in different places at the same time, they will fabricate contradictory memories.

7. *To score high on national aptitude exams, take a test-preparation short course.* Worried that such is true, those that lack such test-training may feel disadvantaged compared to their test-smart competitors. Although crash courses can help people brush up their academic skills and gain tips on how to take the test, studies indicate that they increase Scholastic Aptitude Test scores by a mere 15 points on the average (on the 200 to 800 scale). This small boost is probably a miniature version of the effect of being educated in subjects such as algebra and geometry.

8. *Lie detectors tell the truth.* So President Reagan believed when in 1983 he ordered government employees to take lie detector (polygraph) tests during investigations of leaks and when in 1986 he authorized expanded federal use of polygraph tests. So corporate America believes, when using the polygraph to screen more than a million potential employees annually. Although better than a coin toss—and therefore perhaps useful in criminal investigations (by testing a suspect's emotional response to details known only to the culprit)—lie detector tests err about one-third of the time, usually by declaring the innocent guilty. The truth, if one can believe the near consensus among researchers, is that lie detectors too often lie for their results to be admissible in courts or justifiable for personnel screening.

9. *Projective tests, especially the Rorschach inkblots, provide a sort of psychological X-ray, revealing our hidden traits and impulses.* Actually, projective tests tend not be be very reliable or valid. The most researched of these, the Rorschach, tends to be interpreted differently by different examiners and is not very successful at predicting behavior (such as who is going to commit suicide) or at discriminating between groups (such as heterosexuals and homosexuals). In *The Eighth Mental Measurements Yearbook*, Rolf Peterson summed up fifty years of research and five thousand articles

and books by lamenting that "the general lack of predictive validity for the Rorschach raises serious questions about its continued use in clinical practice."

10. *By discerning the positions of the stars at the time of one's birth, astrologers can analyze character and predict a person's future.* Given that two-thirds of Americans read their horoscope at least occasionally and that almost 40 percent believe astrology has some scientific credence, one can understand why there are now nearly as many astrologers in the Western world as there are psychologists, as many books on astrology as there are on astronomy, and many more magazines devoted to astrology than there are to psychology and astronomy combined. The reasoned arguments against astrology—astrological signs no longer bear any relation to the changing constellations, astrology is naively earth-centered, there is no known way it could work—do not dissuade believers, who are sure it *does* work. But does it? Psychology's contribution has been repeatedly to reveal that it does not—that birth dates are uncorrelated with character traits and that, given someone's birth date, astrologers cannot surpass chance when asked to identify the person from a short line-up of different personality descriptions. Moreover, people cannot pick out their own horoscope from a line-up of horoscopes—hardly surprising given the disagreements among astrologers.

11. *Trained and experienced clinical psychologists can predict future behavior.* After conducting in-depth interviews, they may therefore be trusted to advise which potential parolees can be safely released, which clients are likely to commit suicide, which applicant would be the effective executive. Actually, when such clinical intuition has been put to the test it has almost invariably been found wanting. Researchers Anne Locksley and Charles Stangor explain: "Almost 30 years of . . . research has yet to find a single case" in which prediction by clinical intuition has outperformed simple statistical prediction. There is, however, a very simple rule for predicting people's behavior: observe or ascertain their past behaviors in similar situations. The rule is inexact, but it's the best we have.

12. *Psychological disorders are cured by professional psychotherapy.* So suggest America's advice givers, from Ann Landers to Dr. Ruth, and so believe the millions of people who annually seek mental health services, most of whom report being satisfied with the help they received. Indeed, most depressed or otherwise troubled people improve markedly while in therapy. But so do most such people who go untreated. Moreover, the hundreds of research studies on psychotherapy effectiveness reveal that it generally does not much matter what type of therapy is practiced or how experienced the therapist is. One analysis of thirty-nine studies even revealed no difference between the helpfulness of professional clinicians and of laypeople who were given but a few hours training in empathic listening. For certain specific disorders, such as phobias and certain sexual dysfunctions, there do exist effective treatments, and for broader disorders, such as depression, those who receive certain types of therapy do often improve more than their untreated counterparts. Still, it is a tribute to human resilience that so many disordered people improve without psychotherapy, and it is equally a tribute to our capacity to care for one another that the hope and empathy typically found in therapy can sometimes also be found in close relationships.

Believe It or Not

By clearing the decks of pseudoscientific myths, we prepare ourselves to appreciate genuine wonders. Each new wave of research reminds us of the incompleteness of our prior understandings and hints at the potential foolishness of our current wisdom. Shakespeare's Hamlet seems ever apropos in declaring, "There are more things in heaven and earth, Horatio, than are dreamt of in your philosophy." Who but a few decades ago would have guessed that:

1. *Nerve cells communicate with one another via chemical messengers; these "neurotransmitters" are essential for the brain activities that underly our moods, memories, and mental abilities.* The supplies of specific neurotransmitters influence sleep and arousal, de-

pression and elation, pain and comfort. Hopes are therefore growing that it may become possible to identify and alleviate (through drugs, diet, or even brain tissue transplants) the chemical abnormalities that underly psychological abnormalities.

2. *Massive losses of brain tissue early in life may have minimal long-term effects.* The brain seems more "plastic" than we formerly supposed. If one area is damaged, other areas may gradually take over its functions as undamaged neurons make new connections. In the preschool years, before language and other functions become fixed in specific areas of the brain, even the removal of an entire hemisphere—leaving half a skull filled with fluid—may not prevent the child's becoming a successful and intelligent (although partially paralyzed) adult.

3. *Newborn infants come equipped with remarkable perceptual abilities.* Far from being functionally blind and nondiscriminating, as was formerly supposed, babies come wired for social interaction. Armed with sophisticated eye-tracking machines, pacifiers wired to electronic gear, and other such wonders, researchers have discovered that newborns will turn their heads toward human voices but not other sounds; that they will gaze at a bull's eye pattern nine inches away, which—wonder of wonders—just happens to be the approximate distance of a nursing mother's eyes; that within days of birth they prefer the odor and voice of their mother to that of a stranger; and that they even have a rudimentary ability to copy facial expressions. In psychology textbooks, the mindless newborn has been transformed into the competent infant.

4. *On average, the personalities of any two children from the same family are virtually as different as any two randomly selected children.* In case you wonder whether you read that right, let's try it again in different words: children reared in the same family, whether by their biological or adoptive parents, have personalities hardly more alike than any pair of children selected at random. Identical twins are the only exception, although even they are not so similar in personality as they are in intelligence. It is sometimes said that all parents are environmentalists, until they have their

second child. Actually, report researchers Robert Plomin and Denise Daniels, a family's environment is experienced differently by different children: "the effective environments of siblings are hardly any more similar than are the environments of strangers who grow up in different families." This may partly be due to temperamental differences between children that trigger parents to relate differently to one child than to another and that trigger children to react differently to similar parental demands. (Notice how in hindsight we can begin to transform a finding that has stunned psychologists into tomorrow's common sense.)

5. *Chimpanzees and gorillas exhibit a modest capacity for language.* Washoe, the grandmother of the talking chimps, was taught American sign language, more than fifty words of which she has in turn taught to her foster son, Loulis. Other apes have been taught to communicate with their caretakers or with one another by punching symbols that a computer translates into English. Although their vocabularies and grammar are no better than those of a two- or three-year-old child, they do occasionally combine words creatively, as when describing a swan as a "water bird" or a Pinocchio doll as an "elephant baby."

6. *The things we experience and learn while feeling sad, joyful, angry, or fearful are most easily recalled when we are again in the same mood.* More generally, what is learned in one state (say, while one is drunk or high) is best recalled when again in that state. This mood-memory phenomenon helps explain why moods persist. When happy, we tend to recall happy events, which helps prolong the good mood. When depressed, we tend to recall depressing events, which in turn feeds depressing thoughts. In experiments, people put in a bad mood think more negative thoughts and even interpret a videotape of themselves more negatively. The irony is that *knowing* that our moods color our perceptions of the world doesn't eliminate the effect; we go right on attributing our positive or negative thoughts to reality rather than to our temporary mood.

7. *Memory storage involves the brain's synapses and their neurotransmitters.* For decades neuropsychologists have been searching

for a physical memory record in the brain. They have tried, with little success, surgically to cut memories out of the brain, to erase memory by switching off the brain's electrical activity, and to isolate memories in RNA molecules. Now it seems that experience does leave a physical trace in the brain, but at the synapses—the sites where nerve cells communicate with one another through their neurotransmitters. With experience, new neural connections form and old ones may be eliminated. As a sea snail becomes conditioned to associate water movement with electric shock, its involved synapses also release increased amounts of the neurotransmitter serotonin, making them more efficient at transmitting information. When the formation of the neurotransmitter is blocked with a drug, memory formation is disrupted as well. Although most of what there is to know about memory storage is yet to be learned, hopes grow that new insights into the biology of memory may someday enable treatments of various memory disorders.

8. *Prolonged stress renders people more vulnerable to physical illness.* Although no one needs to be told that stress can trigger butterflies, trips to the bathroom, splitting headaches, or even ulcers, the evidence grows—although not quite so unambiguously as some news reports would suggest—that stress is linked with disease. Under stress, the bodies of reactive, anger-prone people secrete more of the stress hormones that are believed to accelerate the build up of plaques on the arteries of the heart, thereby increasing the risk of a heart attack. The evidence seems even clearer that stress can suppress the immune system, leaving a person more vulnerable to infections and malignancy. It's not that stress *causes* a disease such as cancer, but rather that stress may hinder our disease-fighting capacity.

9. *Electroconvulsive therapy is a mysteriously effective treatment for severe depression.* It sounds barbaric, and it was in the form introduced in 1938: the wide-awake patient would be strapped to a table and jolted with one hundred volts of electricity to the brain, producing bone-rattling convulsions and a temporary loss of consciousness. Today, the patient is first given a general anesthetic and

a muscle relaxant and so awakens thirty minutes later with no memory of the treatment. A 1985 panel of the National Institutes of Health reports that after three such treatments a week for two to four weeks, people who have not responded to drug therapy often improve dramatically. No one knows for sure why. The shock may trigger release of the mood-boosting neurotransmitter norepinephrine, or it may inhibit the overactivity of certain brain areas. Despite its Frankenstein-like image and continuing controversy over its use, electroshock therapy is resurging as more and more psychiatrists view it as a lesser evil than the misery, anguish, and risk of suicide that plagues those with intractable depression.

10. *Rape and other acts of sexual coercion are far more common than suggested by crime rate statistics.* In 1985, 87,000 American females—fewer than one in 1,000—reported being raped. Surveys of women indicate that unreported rapes outnumber those reported by a 10 to 1 ratio. One survey of San Francisco households found that 44 percent of women had experienced at least one attempted or completed rape, only 8 percent of whom reported the incident to police. A study of college women found that 37 percent had experienced the legal definition of rape or attempted rape, 4 percent of whom reported the assault to police. Even larger numbers of women—more than half—have experienced "offensive and displeasing" sexual coercion or harrassment, often while on dates. Surveys of men confirm the women's reports. When "normal" university males are asked whether there is any likelihood they would rape a woman "if you could be assured that no one would know and that you could in no way be punished," about 35 percent admit at least a possibility of their doing so. About one-fourth of college men admit having attempted sexual intercourse using force.

11. *To provide a remarkably accurate snapshot of the opinions of 175 million adult Americans, one need randomly sample fewer than 1 in every 100,000 persons.* Although people commonly express distrust in national surveys because "they've never asked my opinion," the odds that one would be asked aren't much better than those of winning a state lottery. To reemphasize a point already

made, the average American woman is at least 100 and perhaps 1,000 times more likely to be raped this year than she is to be included in the next Gallup presidential preference poll. Nevertheless, if it draws a genuine random sample of merely 1500 people in any country, a survey organization can be 95 percent confident that the survey result will predict to within 3 percent the result that would have been obtained had all the people been surveyed. That simple fact has enabled the Gallup organization to predict the outcomes of the last eighteen national elections in the U.S. with an average error of but 1.4 percent.

12. *To a striking degree, the misperceptions of those in conflict are mutual.* As we see our enemy, so they see us, a phenomenon that social psychologists know as "mirror-image perceptions." Studies of U.S. and Soviet politicians and political statements reveal that people in both nations (a) wish for mutual disarmament; (b) do not wish to disarm while the other side arms; and (c) perceive the other side as seeking military superiority. Thus despite their mutual preference for disarmament ("our nuclear weapons are only for defensive purposes"), both sides arm themselves to counter the other's weapons buildup ("they propose arms control initiatives only for propaganda purposes"). Likewise, both Arabs and Israelis have claimed to be motivated by self-protection but perceive the other side as wanting to obliterate them. And conflicting business executives commonly describe their own behavior as cooperative, the other person's as competitive and demanding. Unfortunately, such diabolical images tend to be self-confirming. John hears a rumor that Mary is saying nasty things about him, so he treats her coldly; noticing such, she badmouths him, confirming his perception. Each party's misperception triggers behavior that reinforces the misperception, creating a vicious circle of conflict.

The abundance of apparent sense and nonsense in the sea of contemporary ideas spurs us to be, as Jesus advised, "wise as serpents and innocent as doves"—innocent as doves in exploring the human creature with wide-eyed wonder, and discerning as serpents in winnowing sense from nonsense. In this process of open yet skeptical scrutiny, Christians and non-Christians alike can benefit from careful scholarship. One cannot read the work of great scholars, noted John Calvin,

"without great admiration. We marvel at them because we are compelled to recognize how preeminent they are. . . . If the Lord has willed that we be helped in physics, dialectic mathematics, and other like disciplines . . . let us use this assistance. For if we neglect God's gift freely offered in these arts, we ought to suffer just punishment for our sloths."

For Further Reading

The Skeptical Inquirer, Box 229, Buffalo, N.Y. 14215-0229.
A quarterly periodical that, with biting humor, sifts sense from nonsense. Recent topics include astrology, "moon madness," animal language, ESP, faith healing, firewalking, ancient astronauts, UFOs, and biorythyms.

Chapter 30

THE PSYCHOLOGY OF RELIGION

Religious ideas are . . . illusions, fulfilments of the oldest, strongest and most urgent wishes of mankind. The secret of their strength lies in the strength of those wishes. As we already know, the terrifying impression of helplessness in childhood aroused the need for protection—for protection through love—which was provided by the father; and the recognition that this helplessness lasts throughout life made it necessary to cling to the existence of a father, but this time a more powerful one. Thus the benevolent rule of a divine Providence allays our fear of the dangers of life; the establishment of a moral world-order ensures the fulfilment of the demands of justice, which have so often remained unfulfilled in human civilization; and the prolongation of earthly existence in a future life provides the local and temporal framework in which these wish-fulfilments shall take place.

SIGMUND FREUD,
THE FUTURE OF AN ILLUSION, 1927

Freud's argument that religion is psychological wish fulfillment offends many people's religious sensibilities. And he is not the only offender. Other theorists have maintained that religion is a sociological phenomenon—a belief system that is socialized by one's family and culture. Still others, such as Karl Marx, have viewed religion in socioeconomic terms—as an opiate for the oppressed masses, who without the hope of pie-in-the-sky-by-and-by would despair of compensation for their suffering. Some even explain religion in evolutionary terms; sociobiologist E. O. Wilson boasts, "We have come to the crucial stage in the history of biology when religion itself is subject to the explanations of the natural sciences. . . . Theology is not likely to survive as an independent intellectual discipline."

In addition to these theoretical efforts to explain the religious impulse there now exists a growing body of research on the psychology of religion. Psychology has made great strides in understanding many other important aspects of human experience—sleep, sex, and hunger—so why not investigate religious behavior as well? What psychological functions does religion serve? What types of people, under what circumstances, are most likely to experience a religious conversion? How might the ways in which people form beliefs influence the formation of their religious ideas? Enough has been learned that in 1988 the prestigious *Annual Review of Psychology* is publishing a full-scale review of the accumulating evidence on the psychology of religion.

As the tide of science rises ever higher and now begins to engulf even religion itself, many Christians see themselves as huddling on a narrowing strip of beach. They build dams and dikes to restrain the tide, but in one spot or another it breaks through. No longer can we feel so sure that "Humans alone have a capacity for language!" that "They'll never be able to explain mental states, such as memories, in terms of brain chemistry!" or that "Phenomena such as ESP, faith healing, and tongues speaking are proof of supernatural forces that disrupt natural processes!"

Imagine, on the other hand, what it would mean to adopt the radically different Christian attitude toward science introduced in Chapter 1, a view that associates divine revelation not with the beach—a shrinking place for God that must yet be defended—but with the tide. Imagine that science, insofar as it seeks and reveals truth, is a gift from God, a vehicle of revelation that coexists with truths revealed in Scripture and interpreted by scholars. Imagine that God's transcendent power, far from being a *separate* psychological force—something extra to be tacked on at the end of a list of other causal factors—operates within a creation that is ordered and upheld by divine providence. Imagine that God's activity is not merely an occasional intervention in a universe from which God is otherwise remote, but rather something more basic. Imagine that, as physicist William Pollard has suggested, "The chances and accidents of history [are] the very warp and woof of the fabric of providence which God is ever weaving." Imagine therefore that events that at one level are intelligible in natural terms—at which level there is no need of a "God hypothesis"—can simultaneously be understood as having spiritual meaning, as God's action in the here and now. Imagine that falling in love, having a child, or achieving good fortune could simultaneously be understood as entirely the outcome of natural processes and also as God's loving handiwork.

What then? What if even our faith commitment itself became understandable as the outcome of natural processes? Would a successful psychology of religion explain away religion, precisely as E. O. Wilson suggested a successful sociobiology of religion would?

The answer is clearly no: *explaining a belief does not explain it away. The truth of a belief is logically distinct from its psychological function.* To see why, imagine that at some point in the future we were to achieve a complete understanding of why some people are believers in God. The psychology of religion has finished its business and is ready to close up shop. Imagine also that we have by that time come to a full understanding of why some people are atheists. It is no chore at all to imagine explanations of atheism that parallel those proposed to explain theism: perhaps, for example, atheism is psychological wish fulfillment by prideful humans who lack the intellectual humility necessary to acknowledge a being infinitely greater than themselves; perhaps for others atheism has socioeconomic motivations—a resistance to believing religion's teachings regarding the value of all human life and the claims of the poor upon one's own material possessions, which are really not possessions but gifts held in stewardship. One could imagine, then, a busy cluster of researchers studying "the psychology of unbelief" (the actual title of a book published some years ago). One can even envision a day when, paraphrasing E. O. Wilson, someone might say that atheism itself is subject to the explanations of the natural sciences and is therefore not likely to survive as a credible intellectual idea.

But hold it (here we rise to the defense of atheism as well as theism): if both belief systems are explained, as would happen in a complete psychology, that cannot mean they are both false. Either God exists or God does not exist, so one of these beliefs must be true.

The point can be extended to any other belief or attitude: knowing why you believe something says nothing about its truth or falsity. To know why someone does or does not believe in extraterrestial beings does not tell us whether such beings exist. To know how someone has been persuaded that a vegetarian diet is healthiest and how someone else has been persuaded that a nonvegetarian diet is healthiest does not decide which diet is, in fact, healthiest. To know the biological, psychological, sociological, and economic factors that lead one person to become a devout believer and another a nonbeliever does not tell us whether God, in fact, exists.

To repeat, the truth of a belief is a logically distinct issue from why anyone believes or does not. So let no one say to you, and never say to anyone, "Your beliefs are not to be taken seriously because you only subscribe to them for such and such reasons." Archbishop William Temple recognized the logical difference between explaining and explaining away when challenged after an address at Oxford: "Well, of course, Archbishop, the point is that you believe what you believe because of the way you were brought up." To which the Archbishop coolly replied, "That is as it may be. But the fact remains that you believe that I believe what I believe because of the way I was brought up, because of the way you were brought up."

The corollary to this idea that explaining religion explains it away is expressed by those who contend that one ought not indoctrinate a child. To do so successfully makes the child's belief system relative to those of the indoctrinators. The assumption here is that if one is socialized or persuaded into a belief, the belief must have less integrity or credibility.

It is true that attitudes and beliefs formed with some degree of choice are more likely to endure and to influence behavior. But is it the case that a self-invented belief has more credibility? And does anyone really *invent* a belief system, free of social influence? B. F.

Skinner has contended that the refusal to modify people's behavior by a planned use of reinforcers does not really increase people's freedom and dignity; rather, it abandons them to the whims of other controls, for better or, more often, for worse. Likewise, parents who are reluctant to impose their values on their children are in effect conceding their children, for better or for worse, to competing influences—television, music, and peers. In truth, however, few parents retreat from persuasive efforts on matters they really care about. We do not leave it to our children to decide for themselves (or under the influence of other persuasive forces) whether it is better to be honest or deceitful, kind or cruel, helpful or hostile. On such matters we all, unapologetically if sometimes ineffectively, do everything we can to socialize our children into what we believe to be right. It is primarily on matters that we ourselves don't really care that much about that we leave our children to work things out for themselves.

Another matter: should it really be any embarrassment if Freud was right that religion serves psychological functions for believers? Or might it not be a greater embarrassment if Freud should be proven wrong and religion found to serve *no* psychological function? If God is, as Christians believe, *for us*, then ought we not expect that worship of God would have secondary benefits (at least insofar as the worship is focused on God rather than on the benefits)?

There is yet another reason why we need not fear the scientific study of religion, or the scientific study of anything else. Yes, we are wise to be wary of hidden values and biases that sneak into scientific interpretation; we do well to be conscious of the limits of scientific explanations and open to other complementary perspectives; we ought not forget that any principle may later be replaced by a deeper principle that incorporates it. Nevertheless, we welcome every advance toward truth, every refutation of falsehood, every approximation of reality that is more faithful to the known data. We do so believing that all truth is God's truth and that our ultimate allegiance is to God alone rather than to the dictates of tradition or human authority.

So let us not resist the tide but welcome it. To the extent that disciplined psychological inquiry covers the sands of ignorance and falsehood with truth it is God's tide. Let us therefore submit ourselves to it—indeed to all of God's revelation, in whatever forms it comes. Let us be assured that God is the author of whatever truth is revealed, however surprising or unsettling it might be. When Freud and Marx argued that religious belief can mask and fulfill hidden self-interest they emphasized a point made by Jesus and the prophets long before them. Let us remember that it is therefore not just our right to freely pursue truth, but our religious duty. It is, indeed, part of what it means to worship God not only with hearts but our minds. Many are those who have committed their hearts to Jesus; fewer are those who have also committed—through disciplined study and inquiry—their minds.

And lest we ever be tempted to think that any of our intellectual achievements are Truth with a capital T, let us finally be conscious of the finiteness of our own feeble efforts to comprehend nature when measured against the infinite wisdom of God. As Agnes Clerke reflected in her century-old A *Popular History of Astronomy*, we know that

What has been done is little—scarcely a beginning; yet it is much in comparison with the total blank of a century past. And our knowledge will, we are easily persuaded, appear in turn the merest ignorance to those who come after us. Yet it is not to be despised, since by it we reach up groping to touch the hem of the garment of the Most High.

For Further Reading

Paloutzian, R. F. *Invitation to the Psychology of Religion*. Glenview, Ill.: Scott, Foresman, 1983.
An introduction to theory and research on the psychology of religious development, conversion, and religious attitudes.

Spilka, B., R. W. Hood, Jr., and R. L. Gorsuch. *The Psychology of Religion: An Empirical Approach*. Englewood Cliffs, N.J.: Prentice-Hall, 1985.

A thorough review of what we know, or think we know, about the function and consequences of religion for children and adults.

NOTES

1: Lessons from the Past: Science and Christian Faith

2 History of American Colleges—B. Spilka, "Religion and Science in Early American Psychology," Presidential Address, American Psychological Association Division, August 1986.

3 "of dust from . . ." Genesis 2:7, RSV.

3 "What is man . . ." Psalm 8:4–5, RSV.

2: Levels of Explanation

6 "Why should I . . ." L. Tolstoy, *My Confessions* (Boston: Dana Estes, 1904), 21.

3: Should There Be a Christian Psychology?

11–14 For surveys of psychological and political views, see C. G. McClintock, C. B. Spaulding, and H. A. Turner, "Political Orientations of Academically Affiliated Psychologists," *American Psychologist* 20 (1965): 211–21; *Chronicle of Higher Education* (April 6, 1970); C. Regan, H. N. Malony, and B. Beit-Hallahmi, "Psychologists and Religion: Professional Factors Associated with Personal Belief" (Paper presented at the American Psychological Association convention, 1976).

13 Mental health workers survey by J. Jensen is reported by A. E. Bergin, "Three Contributions of a Spiritual Perspective to Counseling, Psychotherapy and Behavior Change," in *Counseling and Values*, forthcoming.

14 "of an articulate . . ." R. A. Shweder, "Liberalism as Destiny," *Contemporary Psychology* 20 (1982): 421–24.

14 C. Gilligan, *In a Different Voice: Psychological Theory and Women's Development.* Cambridge, Mass.: Harvard University Press, 1982.

15 "is to 'tell . . ." D. M. MacKay, *Journal of the American Scientific Affiliation* (December 1984).
16 "Christianity has not . . ." C. S. Lewis, *Mere Christianity* (New York: Macmillan, 1960) Book 3, chap. 3.

4: The Brain-Mind Connection

21 "fundamental changes in . . ." D. Hubel, "The Brain," *Scientific American* (September 1979), 45–53.
21 "I am lodged . . ." R. Descartes, *The Meditations and Selections from the Principles of René Descartes* (LaSalle, Ill.: Open Court Publishing, 1948), 94.
21 "The mind seems . . ." W. Penfield, *The Mystery of the Mind* (Princeton: Princeton University Press, 1975), 79.
21 "consists of brain . . ." D. O. Hebb, *Essay on Mind* (Hillsdale, N.J.: Erlbaum, 1980).
22 "The living body . . ." M. Mishkin, "Physiological Psychology and the Future: An Optimist's View," paper presented at the American Psychological Association convention, August 26, 1986.
22 "Everything in science . . ." R. Sperry, "Changed Concepts of Brain and Consciousness: Some Value Implications." *Zygon* 20 (1985): 41–57.

5: Biblical Images of Human Nature

24 "For the Hebrew . . ." H. Wheeler Robinson, "Hebrew Psychology," in *The People and the Book*, ed. A. S. Peake (New York: Oxford University Press, 1925), 366.
25 "rightly affirmed . . ." J. Calvin, *Institutes of the Christian Religion*, ed. J. T. McNeill, trans. F. L. Battles (Philadelphia: Westminster Press, 1975), vol. 1, 192, 185.
26–27 "The unhappy rendering . . ." W. Eichrodt, *Theology of the Old Testament* (Philadelphia: Westminster Press, 1967), 2:134.
27 "present your bodies . . ." Romans 12:1, RSV.
27 "Soul, you have . . ." Luke 12:19, RSV.
27 "What kind of . . ." F. Stagg, *Polarities of Man's Existence in Biblical Perspective* (Philadelphia: Westminster Press, 1973), 49–50.
28 "Such an approach . . ." B. Reichenbach, "Life after Death: Possible or Impossible?" *Christian Scholar's Review* 3 (1974): 232–44.
28 "with few exceptions . . ." O. Cullmann, *Immortality of the Soul or Resurrection of the Dead? The Witness of the New Testament* (New York: Macmillan, 1958).
29 "mixture of body . . ." Seneca, "Seneca of a Happy Life," in *Morals*.
30 "fills us with . . ." Socrates, in Plato's *Phaedo*, ed. H. N. Fowler (Cambridge, Mass.: Harvard University Press, 1913), 231.

6: On Living Peaceably with the Mysteries of Faith

Gordon Allport's meditation, "The Quest for Religious Maturity," in *Waiting for the Lord* (New York: Macmillan, 1978) inspired some of the thoughts in this essay.

31–32 "When I was . . ." 1 Corinthians 13:11, RSV.

32 "We try to . . ." Madeline L'Engle, *Walking on Water: Reflections of Faith & Art* (Wheaton, Ill.: Harold Shaw, 1986), 82.

33 "Sometimes, because of . . ." C. S. Lewis, *Screwtape Proposes a Toast* (London: Collins, Fontana Books, 1967), 69.

34 "There are trivial . . ." N. Bohr, quoted by W. McGuire, "The Yin and Yang of Progress in Social Psychology: Seven Koan," *Journal of Personality and Social Psychology* 26 (1973): 446–56.

34 " 'frightful act of . . ." Kierkegaard summary statement by M. Westphal, "Kierkegaard and the Logic of Insanity," *Religious Studies* 7 (1971): 193–211.

7: How Much Credit (and Blame) Do Parents Deserve?

The many research studies alluded to in this essay are discussed further in D. G. Myers, *Psychology* (New York: Worth Publishers, 1986).

38 Studies of twins reported by J. C. Loehlin and R. C. Nichols, *Heredity, Environment, and Personality* (Austin: University of Texas Press, 1976).

38 "In some domains . . ." T. Bouchard, interviewed on *Nova: Twins* (Program broadcast by the Public Broadcasting Service, December 6, 1981).

39 "Our studies suggest . . ." S. Scarr, "What's a Parent to Do?" (Conversation with E. Hall) *Psychology Today* (May 1984), 58–63.

39 "Many of our . . ." J. W. Macfarlane, "Perspectives on Personality Consistency and Change from the Guidance Study," *Vita Humana* 7 (1964): 115–26.

40 "Judge not" Matthew 7:1.

8: The Mystery of the Ordinary

42 Regarding the nonexistence of a reproducible ESP phenomenon, British expert C. E. M. Hansel summarizes, "After a hundred years of research, not a single individual has been found who can demonstrate ESP to the satisfaction of independent investigators." *ESP and Parapsychology: A Critical Reevaluation* (Buffalo, N.Y.: Prometheus Books, 1980), 314. John Beloff, past president of the Parapsychological Association, appears to agree: "No experiment showing the clear existence of the paranormal has been consistently repeated in other laboratories." See "Why Parapsychology Is Still on Trial," *Human Nature* (December 1979), 68–74.

42 "If a prophet . . ." Deuteronomy 18:22, TEV.

42 "I am God . . ." Isaiah 46:9, RSV.

43 "It is a . . ." Sir Arthur Conan Doyle, "A Study in Scarlet," in *The Complete Sherlock Holmes Long Stories* (London: Book Club Associates, 1973), 60; *The Adventures of Sherlock Holmes* (London: Cathay Books, 1983), 39.

44 "I have uttered . . ." Job 42:3, RSV.

47 "The mere existence . . ." L. Thomas, *The Medusa and the Snail* (New York: Viking Press, 1979), 156–57. The earlier description of our language monitoring feat is also adapted from this book, 123–24.

9: On Discerning Sense from Nonsense

49 "Success depends on . . ." C. S. Lewis, *Screwtape Letters* (London: The Centenary Press, 1942), 106, 128–29.

50–53 J. R. Vokey and J. D. Read describe their research and the public controversy over backward masking in "Subliminal Messages: Between the Devil and the Media," *American Psychologist* 40 (1985): 1231–39.

10: Through the Eyes of Faith

54 "He has a . . ." John 10:20–21, RSV.

55 The experiment with the blurred pictures appears in A. G. Greenwald, A. R. Pratkanis, M. R. Leippe, & M. H. Baumgardner, "Under What Conditions Does Theory Obstruct Research Progress," *Psychological Review* 93 (1986): 216–29.

56–57 "What we learn . . ." C. S. Lewis, *Miracles* (New York: Macmillan, 1947), p. 11.

58 "I have wondered . . ." P. Kurtz, *The Transcendental Temptation: A Critique of Religion and the Paranormal.* (Buffalo, N.Y.: Prometheus Books, 1986), 105.

11: The Day of Rest

59 "hopelessly indebted . . ." C. S. Lewis, *God in the Dock* (Glasgow-Collins, Fount Paperback, 1979), 63.

59–60 The description of Suedfeld's research and of the other examples of solitude is drawn from his "Aloneness as a Healing Experience," in *Loneliness: A Sourcebook of Current Theory, Research, and Therapy,* Ed. L. A. Peplau and D. Perlman (New York: Wiley, 1982). For further information see P. Suedfeld, "The Benefits of Boredom: Sensory Deprivation Reconsidered," *American Scientist* 63 (1975): 60–69, and P. Suedfeld, *Restricted Environmental Stimulation: Research and Clinical Applications* (New York: Wiley, 1980).

61 Outward Bound study: H. W. Marsh, G. E. Richards, and J. Barnes, "Multi-dimensional Self-Concepts: The Effect of Participation in an Outward Bound Program," *Journal of Personality and Social Psychology* 50 (1986): 195–204.

62 "finished his work . . ." Genesis 2:2–3.

12: Are We Determined or Free?

63 "The issue is . . ." W. James, "The Dilemma of Determinism," in *The Will to Believe*, ed. W. James (New York: Longmans, Green & Co., 1896), 145–83.

65 "passionless, feelingless association . . ." Colin Turnbull, *The Mountain People* (New York: Simon and Schuster, 1972).

65 Langdon Gilkey, *Shantung Compound* (New York: Harper & Row, 1966).

67 "a man reaps . . ." Galatians 6:7, NEB.

67 "train up a . . ." Proverbs 22:6, RSV.

67 "unto the third . . ." Numbers 14:18, KJV.

68–69 "If we believe . . ." M. Luther, *The Bondage of the Will*, trans. J. I. Packer and O. R. Johnston (Old Tappen, N.J.: Fleming Revell, 1957), 317.

69 "constantly changing his . . ." J. Edwards, *Freedom of the Will*, ed. P. Ramsey (New Haven: Yale University Press, 1957), 253, 27.

69 "Our wills themselves . . ." Augustine, *The City of God*, bk. 5, chap. 9.

69 "Nothing in truth . . ." Luther, *Bondage*, 268.

69 "cannot be retained . . ." J. Calvin, *Institutes of the Christian Religion*, vol II, ed. J. T. McNeill, trans. F. L. Battles (Philadelphia: Westminster Press, 1975), 255.

70 "grace operates . . ." Michael Novak, "Is He Really a Grand Inquisitor?" in *Beyond the Punitive Society*, ed. H. Wheeler (San Francisco: Freeman, 1975), 235.

71 "What I so . . ." C. S. Lewis, *Mere Christianity*, Book IV, ch. 7.

71 "I . . . yet not . . ." 1 Corinthians 15:10, KJV.

13: Memorable Messages

75 Earlier versions of this chapter, coauthored by the memory researcher John Shaughnessy, appeared in the *Church Herald, Christian Ministry, Military Chaplain's Review*, and *The Human Connection*, by M. Bolt and D. Myers (Downers Grove, Ill.: InterVarsity, 1984).

76 "talked about prejudice . . ." T. J. Crawford, "Sermons and Racial Tolerance and the Parish Neighborhood Context," *Journal of Applied Social Psychology* 4, (1974): 1–23.

77 "If those who . . ." W. Strunk and E. B. White, *The Elements of Style*, 3d ed. (New York: Collier Macmillan, 1979), 21.

78 "The procedure is . . ." J. D. Bransford and M. K. Johnson, "Consideration of Some Problems of Comprehension," in *Visual Information Processing*, ed. W. Chase (New York: Academic Press, 1973), 383–483.

78 L. Hasher, D. Goldstein, and T. Toppino, "Frequency and the Conference of Referential Validity," *Journal of Verbal Learning and Verbal Behavior* 16 (1977): 107–12.

80 "No reception without . . ." W. James, *Talks to Teachers on Psychology: And to Students on Some of Life's Ideals* (New York: Holt, 1922), 33.

14: To Err Is Human

84 "Man's greatness lies . . ." Pascal, *Pensees*, 1670 (no. 157).

84 "no one can . . ." Psalms 19:12, TEV. For detailed information on illusory thinking, see D. G. Myers, *The Inflated Self: Human Illusions and the Biblical Call to Hope* (San Francisco: Harper & Row, 1980).

86 "The Bible always . . ." M. Marty, "The Bible in American Culture," *Perspectives* (June, 1987): 4–6.

15: Superstition and Prayer

90 "I felt that . . ." Reported by Beth Spring, "One Step Closer to a Bid for the Oval Office," *Christianity Today* (October 17, 1986), 39–45.

90–91 "General Patton . . ." G. S. Patton, Jr., *War As I Knew It* (Boston: Houghton Mifflin, 1949), pp. 184–85, cited by J. T. Burtchaell, *Philemon's Problem* (Chicago: Acta Foundation, 1973).

91 "acts which are . . ." B. Malinowski, *Magic, Science, and Religion* (Garden City, N.Y.: Doubleday, 1948), 68.

92 Prayer test, S. G. Brush, "The Prayer Test," *American Scientist* 62 (1974): 561–63.

92 "not to put . . ." Matthew 4:7, NEB.

92 "impossibility of empirical . . ." C. S. Lewis, *Miracles* (New York: Macmillan, 1947), 215.

93 "You cannot tell . . ." Luke 17:20–21, NEB.

94 "Our Father which . . ." Matthew 6:9–13, KJV.

94 "The prayer of . . ." J. I. Packer, *Evangelism and the Sovereignty of God* (Downer's Grove, Ill.: InterVarsity, 1961), 11.

95 "What are we . . ." Matthew 6:31–33, NEB.

95 "The Lord is . . ." Philippians 4:6–7, NEB.

16: Watch Your Language

97 "Inexpressible" 2 Corinthians 12:2–4.

98 "Language itself shapes . . ." B. L. Whorf, "Science and Linguistics," in J. B. Carroll, ed., *Language, Thought, and Reality: Selected Writings of Benjamin Lee Whorf* (Cambridge, Mass.: MIT Press, 1956).

100 "one is conscious . . ." W. I. McGuire, C. V. McGuire, P. Child, and T. Fujioka, "Salience of Ethnicity in the Spontaneous Self-Concept as a Function of One's Ethnic Distinctiveness in the Social Environment, *Journal of Personality and Social Psychology* 36 (1978): 511–20.

100–101 "that they may . . ." John 17:21, RSV.

17: You Are Gifted

104 "have the sneaking . . ." D. Sisk, and other ideas from this essay can be found in D. G. Myers and J. Ridl, "Aren't All Children Gifted?" *Today's Education* (February–March 1981), 30GS–34GS. For further critique of the gifted child movement see also D. Weiler, "The Alpha Children: California's Brave New World for the Gifted," *Phi Delta Kappan* (November 1978), 185–87; D. Feldman, "Toward a Nonelitist Conception of Giftedness," *Phi Delta Kappan* 60 (1979), 660–63; A. G. Remley, "All the Best for the Brightest: But What About the Other 95 Percent?" *The Progressive* (May 1981), 44–47; I. J. Hughes, "All Our Children are Gifted," *Presbyterian Survey* (October 1983), 58; and J. Kristofco, "Aren't All Kids Gifted?" *Family Learning* (May/June 1984), 57–59.

104 "not enough gifted . . ." E. Garfield, "Will a Bright Mind Make Its Own Way?" *Current Contents* (December 22, 1980), 5–15. Garfield also provides addresses of national associations and periodicals pertaining to the gifted.

104 "provide opportunities and . . ." J. Gardner, *Excellence* (New York: Norton, 1984).

106 "Having gifts that . . ." Romans 12:6, RSV.

18: To Accept or to Change?

Documentation for weight control and temperament research may be found in D. G. Myers, *Psychology* (New York: Worth Publishers, 1986).

110 R. Plomin, "Behavior Genetics," in *Current Topics in Human Intelligence*: *Research Methodology*, ed. D. K. Detterman (Norwood, N.J.: Ablex, 1984).

111 New hints at biological factors: L. Ellis and M. A. Ames, "Neurohormonal Function and Sexual Orientation: A Theory of Homosexuality—Heterosexuality," *Psychological Bulletin* 101 (1987): 233–58.

112 "We all battled . . ." "Ted," "Ex-gay Testimony: What You Should Know About Ted," *Evangelicals Concerned*, P.O. Box 2167, San Francisco, CA 94126.

112 "I had feelings . . ." Stephanie, in J. Alexander, "We're Living Who We Are," *The Other Side* (June 1978), 27–29.

113 M. Friedman and D. Ulmer, *Treating Type A Behavior—And Your Heart* (New York: Knopf, 1984).

113–114 "The biggest truth . . ." N. Cousins, "The Taxpayers' Revolt: Act Two," *Saturday Review* (September 16, 1978), 56.

115 Colin Cook's resignation was explained in a letter to supporters of Homosexuals Anonymous by Coordinator Dan Roberts (Quoted in *Record*, winter 1987) and reported by R. Frame, "Leading Ex-Gay Figure Resigns Counseling Post," *Christianity Today* [March 6, 1987], 7).

19: And God Said, It Is Very Good

116–117 "were both naked . . ." Genesis 2:25–26, 31, RSV.
117 D. M. Joy, *Rebonding: Preventing and Restoring Damaged Relationships* (Waco, Tex.: Word, 1986).
118 For research on the correlates of cohabitation, premarital sex, and marriage, see M. D. Newcomb and P. M. Bentler, "Marital Breakdown," in *Personal relationships*, vol. 3, *Personal Relationships in Disorder*, ed. S. Duck and R. Gilmour (London: Academic Press, 1981); M. D. Newcomb, "Cohabitation, Marriage and Divorce Among Adolescents and Young Adults," *Journal of Social and Personal Relationships* 3 (1986): 473–94; and M. Argyle and M. Henderson, *The Anatomy of Relationships* (Harmondsworth, Middlesex, England: Penguin Books, 1985).
118 For research on television and pornography, see D. G. Myers, *Psychology* (New York: Worth Publishers, 1986), or *Social Psychology* (New York: McGraw-Hill, 1987).
119 "If we value . . ." D. Zillmann, "Effects of Repeated Exposure to Nonviolent Pornography" (Testimony to the Attorney General's Commission on Pornography, Houston, September 11, 1985).
119 "self-gratification . . ." A. E. Ellis, "Psychotherapy and Atheistic Values: A Response to A. E. Bergin's 'Psychotherapy and religious values,' " *Journal of Consulting and Clinical Psychology*, 48 (1980): 635–39.
120 "promiscuous recreational sex . . ." D. Baumrind, "Adolescent Sexuality: Comment on Willliams' and Silka's Comments on Baumrind," *American Psychologist* 37 (1982): 1402–3.
120 "Sexologists should universally . . ." T. McIlvenna, quoted in *Behavior Today Newsletter* (July 28, 1986), 2.

20: This Way to Happiness

124 For research on the effects of happiness, of self-talk, and of facial expressions, see D. G. Myers, *Psychology* (New York: Worth Publishers, 1986).
124–125 Income-happiness data are summarized in T. Ludwig, M. Westphal, R. Klay, and D. Myers, *Inflation, Poortalk, and the Gospel* (Valley Forge, Pa.: Judson Press, 1981).
125 "When any situation . . ." Freud, quoted by E. Berscheid, "Emotion," in *Close Relationships*, ed. H. H. Kelley, et al. (New York: Freeman, 1983).
125 "While happiness . . ." P. Brickman and D. Campbell, ibid.
125 "No happiness lasts . . ." Seneca, *Agamemnon*, A.D. 60.

126 "Chapter One of . . ." C. S. Lewis, *The Last Battle* (New York: Collier Books, 1974), 184.

126 "all you have . . ." A. Maslow, quoted in B. G. Maslow, ed., *Abraham H. Maslow: A Memorial Volume* (Monterey, Calif.: Brooks/Cole, 1972).

126 Experiment on viewing poverty and tragedy by M. Dermer, et al., "Evaluative Judgements of Aspects of Life as a Function of Vicarious Exposure to Hedonic Extremes," *Journal of Personality and Social Psychology* 37 (1979): 247–60.

126 "The streams of . . ." Benjamin Franklin, *Autobiography*, 1798.

127 "Martha, Martha, you . . ." Luke 10:41, RSV.

127 Religion-happiness survey findings from A. M. Greeley, "The State of the Nation's Happiness," *Psychology Today* (January 1981), pp. 14–16; B. Hunsberger, "Religion, Age, Life Satisfaction, and Perceived Sources of Religiousness: A Study of Older Persons" (Paper presented to the American Psychological Association convention, 1984); G. Gallup, Jr., "Religion in America," *The Gallup Report*, no. 222 (March 1984).

127 "He who loses . . ." Matthew 10:39, RSV. (See also Mark 8:35, Luke 17:33 and John 12:25).

21: A New Look at Pride

130–134 Documentation for research findings may be found in D. G. Myers, *Social Psychology*, 2d ed. (New York: McGraw Hill, 1987).

129 "they despise themselves . . ." C. Rogers, "Reinhold Niebuhr's *The Self and the Dramas of History*: A Criticism" *Pastoral Psychology* 9 (1958): 15–17.

130 "People experience life . . ." A. Greenwald, quoted by D. Coleman, "A Bias Puts Self at Center of Everything," *The New York Times*, June 12 (1984), C1, C4.

131 "The arch-flatterer . . ." Plutarch, *De Adulatio et Amico*, as quoted by Bacon, *Essays: Of Love*.

131–132 "Every moving illustration . . ." M. Yaconelli, "Fantasy Christianity," *The Wittenburg Door* (December/January 1987): 31, 30.

134 "in humility count . . ." Phillippians 2:3, RSV.

135 "Each nation feels . . ." D. Carnegie, *How to Win Friends and Influence People* (New York: Pocket Books, 1936/1964), 102.

22: The Power of Positive Thinking

137 "If any man . . ." Mark 8:34, RSV.

137 Jesus on the value of human life is in Matthew 6:26, 12:12, RSV.

137 "I'm me and . . ." Quoted in A. Hoekema, *The Christian Looks at Himself* (Grand Rapids, Mich.: Eerdmans, 1975), 15.

138–140 Research on self-esteem and positive thinking is documented in D. G. Myers, *Psychology* (New York: Worth Publishers, 1986).

140 "what you believe . . ." B. Smallwood and S. Kilborn, "Success," published by the Prince Corporation, Holland, Michigan.

140 "Those who really . . ." R. DeVos, *Detroit Free Press*, October 24, 1982, 10A.

140 "fix men's affections . . ." C. S. Lewis, *The Screwtape Letters* (London: Geoffrey Bles, The Centenary Press, 1942), 77–78.

141 Experiments on anxiety over failure: J. K. Norem and N. Cantor, "Defensive pessimism: Harnessing anxiety as motivation," *Journal of Personality and Social Psychology* 51 (1986): 1208–17.

141 "The present is . . ." Pascal, *Pensees.*

142 "at the moment . . ." C. S. Lewis, *Screwtape*, 71.

142 "not that we . . ." D. Voskuil, *Mountains into Goldmines: Robert Schuller and the Gospel of Success* (Grand Rapids, Mich.: Eerdmans, 1983), 147–48.

142 "If anyone would . . ." C. S. Lewis, *Mere Christianity* (New York: Macmillan, 1960), 99.

23: Is Christianity Beneficial or Hazardous to Your Mental Health?

146 "obsessional neurosis" S. Freud, *The Future of an Illusion* (Garden City, N.Y.: Doubleday, 1964), 71.

146 G. W. Albee, "Toward a Just Society: Lessons from Observations on the Primary Prevention of Psychopathology," *American Psychologist* 41 (1986): 891–98.

146–147 C. D. Batson and W. L. Ventis, *The Religious Experience: A Social-Psychological Perspective* (New York: Oxford University Press, 1982). See also A. E. Bergin, "Religiosity and Mental Health: A Critical Reevaluation and Meta-Analysis," *Professional Psychology* 14 (1983): 170–84; and A. E. Bergin, K. S. Masters, and P. S. Richards, "Religiousness and Mental Health Reconsidered: A Study of an Intrinsically Religious Sample," *Journal of Counseling Psychology* 34 (1987).

148 "adjusting (or conforming) . . ." R. F. Paloutzian, *Invitation to the Psychology of Religion* (Glenview, Ill.: Scott, Foresman, 1983), 183.

149 "It is easy . . ." R. Niebuhr, *Beyond Tragedy* (New York: Scribner's, 1937), 97.

149–150 "There has not . . ." C. Jung, *Modern Man in Search of a Soul* (New York: Harcourt, 1933), 229.

150 Purpose in life research is summarized in Paloutzian, *Invitation*. Subtance use and meaninglessness in life have been researched by M. D. Newcomb and L. C. Harlow, "Life Events and Substance Use Among Adolescents; Mediating Effects of Perceived Loss of Control and Meaningless in Life," *Journal of Personality* and *Social Psychology*, 51 (1986): 564–77.

150 Social support research is critically evaluated in S. Cohen and S. L. Syme, eds., *Social Support and Health* (Orlando, Fla.: Academic Press, 1985).

150 God-concept and self-concept research is described by B. Spilka, R. W. Hood, Jr., and R. L. Gorsuch, *The Psychology of Religion: An Empirical Approach* (Englewood Cliffs: N.J.: Prentice-Hall, 1985), 310–11.

24: Grace: God's Unconditional Positive Regard

151 "in sin and . . ." S. J. DeVries, "Sin, Sinners," *Interpreter's Dictionary of the Bible* (Nashville: Abingdon Press, 1962), 4:361.

152 "I have seen . . ." Ecclesiastes 1:14, RSV.

152 "Happy are those . . ." Matthew 5:3, TEV.

152 "the only way . . ." J. Jellison, *I'm Sorry I Didn't Mean To and Other Lies We Love to Tell* (New York: Catham Square Press, 1977), 155.

152 "Christian religion is . . ." C. S. Lewis, *Mere Christianity*. New York: Macmillan, 1960, 25.

153 "I no longer . . ." Philippians 3:9, TEV.

153 "To give up . . ." W. James, *Principles of Psychology*, vol. I (Cambridge, Mass.: Harvard University Press, 1981), 296–97.

154 "as persons are . . ." C. R. Rogers, *A Way of Being* (Boston: Houghton Mifflin, 1980), 116.

154 J. E. Dittes, "Justification by Faith and the Experimental Psychologist," *Religion in Life*, 1959, 28, 567–76.

154 "Come to me . . ." Matthew 11:28, RSV.

155 "I am convinced . . ." Romans 8:38–39, NEB.

25: Values in Psychotherapy

156–161 A. E. Bergin, Psychotherapy and Religious Values, *Journal of Consulting and Clinical Psychology* 48 (1980): 95–105. Critiques by G. B. Walls and A. Ellis and a reply by Bergin follow on pages 635 to 645. For further critique and discussion see A. E. Bergin, "Proposed Values for Guiding and Evaluating Counseling and Psychotherapy," *Counseling and Values* 29 (1985): 99–116; and J. F. Curry, "Christian Humanism and Psychotherapy: A Response to Bergin's Antithesis," *Zygon*, forthcoming.

157 "The only question . . ." C. Rogers, quoted by M. A. Wallach and L. Wallach, "How Psychology Sanctions the Cult of the Self," *Washington Monthly* (February 1985), 46–56.

157 "Selfless Syndrome" in J. P. Lemkau and C. Landau, "The 'Selfless Syndrome': Assessment and Treatment Considerations," *Psychotherapy* 23 (1986): 227–33.

159–160 J. D. Gartner, "Antireligious Prejudice in Admissions to Doctoral Programs in Clinical Psychology," *Professional Psychology: Research and Practice* (1986): 473–75.

26: Nice People and Evildoers

For further information on the research in this chapter, see D. G. Myers, *Social Psychology* (New York: McGraw-Hill, 1987).

164 "The safest road . . ." C. S. Lewis, *The Screwtape Letters* (London: Geoffrey Bles, The Centenary Press, 1942), 65.

165 "Over a drink . . ." *Ibid.*, 36–37.

167 "For total greed . . ." L. Thomas, *The Lives of a Cell: Notes of a Biology Watcher* (London: Allen Lane, 1980), 109.

27: Behavior and Attitudes: Action and Faith

Portions of this chapter appeared previously in *Christianity Today* and in *The Human Connection*, by M. Bolt and D. Myers (Downers Grove, Ill.: InterVarsity Press, 1984).

For further information on the attitude-behavior relationship see Chapter 2, "Behavior and Attitudes," in D. G. Myers, *Social Psychology* (New York: McGraw-Hill, 1987).

For further biblical and theological commentary concerning the faith-action relationship see Chapter 6, "Christian Action and Christian Faith," in D. G. Myers, *The Human Puzzle: Psychological Research and Christian Belief* (San Francisco: Harper & Row, 1978).

170 "Many Germans experienced . . ." R. Grunberger, *The 12-Year Reich: A Social History of Nazi Germany 1933–1945* (New York: Holt, Reinhart & Winston, 1971), 27.

171–172 "This feature, more . . ." O. A. Piper, "Knowledge," *Interpreter's Dictionary of the Bible*, vol. 3, 44.

172 "he who does . . ." John 3:21, RSV.

172 "is born of . . ." J. Calvin, *Institutes of the Christian Religion*, vol. I, ed. J. T. McNeill, trans. F. L. Battles (Philadelphia: Westminster Press, 1975), 72.

172 "The proof of . . ." S. Kierkegaard, *For Self-Examination and Judge for Yourselves*, trans. W. Lowrie (Princeton: Princeton University Press, 1944), 88.

172 "Only the doer . . ." K. Barth, quoted by J. H. Westerhoff III, *Values for Tomorrow's Children* (Philadelphia: Pilgrim Press, 1971), 44.

172 "Follow the way . . ." B. Pascal, *Thoughts (Pensees)*, trans. W. F. Trotter, *World Masterpieces*, ed. M. Mack (New York: Norton, 1965), 2:38.

172–173 "Believe in God . . ." C. S. Lewis, *Christian Reflections* (Glasglow: Collins, Fount Paperbacks, 1981), 61, and *Mere Christianity*, Book III, Ch. 9. (New York: Macmillan, 1960).

28: The Fruit of the Spirit

176 "You shall love . . ." Matthew 22:39, RSV.

176 "Love your enemies . . ." Matthew 5:44, RSV.

176 Reviews of the religion-morality literature may be found in B. Spilka, R. W. Hood, Jr., and R. L. Gorsuch, *The Psychology of Religion: An Empirical Approach* (Englewood Cliffs, N.J.: Prentice-Hall, 1985); and in C. D. Batson and W. L. Ventis, *The Religious Experience: A Social-Psychological Perspective* (New York: Oxford University Press, 1982).

177 "Christians have given . . ." M. L'Engle, *Walking on Water: Reflections on Faith and Art* (Wheaton, Ill.: Harold Shaw, 1980), 59.
178 "There is no . . ." C. D. Batson, P. A. Schoenrade, and V. Pych, "Brotherly Love or Self-Concern? Behavioral Consequences of Religion," in *Advances in the Psychology of Religion*, ed. L. D. Brown (Oxford: Pergamon Press, 1985).
178 Prejudice-religion studies are reviewed in D. G. Myers, *Social Psychology* (New York: McGraw-Hill, 1987).
178 The St. Andrews reign of terror is described in J. K. Robertson, *About St. Andrews—and About* (St. Andrews: J. & G. Innes, 1973).
178 "Not everyone who . . ." Matthew 7:21, RSV.
178 "Men never do . . ." B. Pascal, *Pensées*, No. 895.
178 On contributions money, see Spilka, et al., *Psychology of Religion*; on volunteerism, see P. L. Benson, et al., "Intrapersonal Correlates of Nonspontaneous Helping Behavior," *Journal of Social Psychology* 110 (1980): 87–95.
178 Gallup survey in "Religion in America," *The Gallup Report*, no. 222 (March 1984).
179 "It makes prejudice . . ." G. W. Allport, *The Nature of Prejudice* (New York: Doubleday, Anchor Books, 1958), 413.
179 "We have just . . ." J. Swift, *Thoughts on Various Subjects*, 1814.
179–180 The Eric Liddell story is assembled from a biography by S. Magnusson, *The Flying Scotsman* (London: Quartet Books, 1981), from David J. Michell's "I Remember Eric Liddell," in *The Disciplines of the Christian Life*, ed. E. Liddell (London: Triangle, 1985), and from L. Gilkey's *Shantung Compound* (New York: Harper & Row, 1966).
181 "literally, God-controlled . . ." A. P. Cullen, quoted by S. Magnusson, *The Flying Scotsman*, op cit, 175–76.

29: More Sense and Nonsense

Most of the research findings described in this essay are documented and described at greater length in D. G. Myers, *Psychology* (New York: Worth Publishers, 1986).
183 "If a prophet . . ." Deuteronomy 18:22, TEV.
184 "Gentlemen, I would . . ." T. Jefferson, quoted by Persi Diaconis, "Statistical Problems in ESP Research," *Science* 201 (1978): 131–35.
186 Sleepwalking, sleeptalking findings are from W. Dement, *Some Must Watch While Some Must Sleep* (New York: Norton, 1978).
186 Contradictory "past lives" of hypnotized subjects are reported by R. L. Atkinson, R. C. Atkinson, E. E. Smith, and E. R. Hilgard, *Introduction to Psychology* (San Diego: Harcourt Brace Jovanovich, 1987), 143.
188 "the general lack . . ." R. Peterson, "Review of the Rorschach." In *The Eighth Mental Measurements Yearbook*, vol. 1, ed. O. K. Buros (Highland Park, N.J.: Gryphon Press, 1978).
188 Astrology beliefs are reported by Kendrick Frazier, "Survey Examines Level of Scientific Illiteracy in U.S.," *The Skeptical Inquirer* 11 (1986–87): 116–19. Number of astrologers and astrology magazines is reported by G. Dean, "Does

Astrology Need to Be True? Part 1: A Look at the Real Thing," *The Skeptical Inquirer* 11 (1986–87): 166–84. For a sample test of astrology, see S. Carlson, "A Double-Blind Test of Astrology," *Nature*, 318 (1985): 419–25.

188 "Almost 30 years . . ." Anne Locksley and Charles Stangor, "Why Versus How Often: Causal Reasoning and the Incidence of Judgmental Bias," *Journal of Experimental Social Psychology* 20 (1984): 470–83.

191 "The effective environments . . ." R. Plomin & D. Daniels, "Why are Children in the Same Family so Different from one Another?" *Behavioral and Brain Sciences* 60 (1987): 1–60. This article includes 31 commentaries, most of which accept the lack of child-to-child personality similarity within families but debate the reasons for it.

193 Sexual coercion findings are summarized in E. R. Allgeier, "Coercive Versus Consensual Sexual Interactions" (G. Stanley Hall Lecture presented to the American Psychological Association, Washington, D.C., 1986).

194 For recent findings on Soviet-American mirror-image perceptions, see S. Plous, "Perceptual Illusions and Military Realities: A Social-Psychological Analysis of the Nuclear Arms Race," *Journal of Conflict Resolution* 29 (1985): 363–89.

194 "Wise as serpents . . ." Matthew 10:16, RSV.

195 "without great admiration . . ." J. Calvin, *Institutes of the Christian Religion*, vol. II, ed. J. T. McNeill, trans. F. L. Battles (Philadelphia: Westminster Press, 1975), 274–75.

30: The Psychology of Religion

196 "We have come . . ." E. O. Wilson, *On Human Nature* (Cambridge, Mass.: Harvard University Press, 1978), 192.

197 The image of Christians huddling against the rising tide of science is adapted from C. S. Lewis, *Screwtape Proposes a Toast* (London: Collins, Fount Paperbacks, 1965), 54.

197 "The chances and . . ." W. Pollard, *Chance and Providence* (New York: Harper & Brothers, 1958), 71.

199 "That is as . . ." Archbishop W. Temple, quoted in M. A. Jeeves, *Psychology and Christianity: The View Both Ways* (Leicester, England: InterVarsity Press, 1976), 133.

201 "What has been . . ." Agnes Clerke, *A Popular History of Astronomy During the Nineteenth Century* (Edinburgh: A. & C. Black, 1885).

INDEX

Brain (continued)
20; memory and, 191–92; mind
and, 21–23; mysteries of, 43–47;
neurotransmitters, 20
Brand, P., 48
Bransford, J. D., 78, 207
Brickman, P., 125, 210
Broca, P., 20
Brown, L. D., 215
Browning, E. B., 131
Brush, S. G., 208
Buchan, A., 182
Buchanan, D., 150
Buddha, 61
Buddhism, Z., 61
Bufford, R., 73
Buros, O. K., 215
Burtchaell, J. T., 208
Buttrick, G. A., 96

Calvin, J., 25, 68, 69, 129, 172, 195,
204, 207, 214, 216
Campbell, D., 125, 210
Camus, A., 88
Cantor, N., 212
Carlson, S., 216
Carnegie, D., 135, 211
Carpenter, N., 1
Carroll, J. B., 208
Carroll, L., 51
Carter, J., 141
Chance encounters, 39
Charismatics, 173
Chess, S., 40
Chesterfield, Lord, 83
Child, P., 209
Child-rearing, 13, 36–40
Children's sermons, 32
Chimpanzees, 191
Christian education, 174
Christian nurture, 174
Christian psychology, 11–17
Cialdini, R. B., 81
Clerke, A., 201, 216
Clinical psychology, 156–61, 188
Cognitive behavior therapy, 139
Cognitive conceit, 85, 132
Cohen, S., 212
Coleman, D., 211
College Board, 130, 187

Competition, 166–67
Conception, 46
Confirmation bias, 85
Consciousness, 9, 21–22, 44–45, 51,
61
Cook, C., 115, 209
Cooper, J., 30
Cosgrove, M., 23, 73
Cousins, N., 144, 209
Crawford, T. J., 75, 207
Cromwell, O., 87
Cullen, A. P., 215
Cullmann, O., 28, 204
Cults, 42, 147
Curry, J. F., 213

Daniels, D., 191, 216
Dante, 77
Darwin, C., 1
Dean, G., 215
Death, 24, 28–29, 148; fear of, 150,
185
De Beauvoir, S., 11
Dement, W., 215
Demon possession, 41
Depression, 20, 38, 44, 133, 149,
189, 190; electroconvulsive therapy,
192–93
Dermer, M., 211
Descartes, R., 21, 204
Determinism, 63–72
Detterman, D. K., 209
Development, 46–47
DeVos, R., 140, 212
DeVries, S. J., 151, 213
Diaconis, P., 215
Dieting, 110, 114
Dittes, J., 154, 213
Divine providence, 197
Divine sovereignty, 33, 69
Doubt, 88
Doyle, A. C., Sir, 41, 63, 206
Dreams, 185
Drug abuse, 177
Dualism, 21, 23, 24, 98
Duck, S., 210

Education, 14
Edwards, J., 68, 69, 71, 207
Eichrodt, W., 26, 204